Primer on Automotive Lightweighting Technologies

Primer on Automotive Lightweighting Technologies

Authored by

RAGHU ECHEMPATI

CRC Press
Taylor & Francis Group
Boca Raton London New York

CRC Press is an imprint of the
Taylor & Francis Group, an **informa** business

First edition published 2021
by CRC Press
6000 Broken Sound Parkway NW, Suite 300, Boca Raton, FL 33487-2742
and by CRC Press

2 Park Square, Milton Park, Abingdon, Oxon, OX14 4RN

Library of Congress Cataloging-in-Publication Data
Names: Echempati, Raghu, author.
Title: Primer on automotive lightweighting technologies / authored by Raghu Echempati.
Description: First edition. | Boca Raton : CRC Press, 2021. | Includes bibliographical references and index. | Summary: "Aluminum is replacing steel for its strength to weight ratio, equal or better stiffness and toughness properties, durability and manufacturability considerations. The book introduces basic ideas and principles of designing engineering components with aluminum. Topics include application of the knowledge to understand how automotive body and structures are designed, as well as other major and smaller automotive components, such as engine blocks and its components, chassis systems, and wheels, to name a few. Applications of aluminum in other fields such as aircraft and off-road vehicles are also covered"-- Provided by publisher.
Identifiers: LCCN 2020043772 (print) | LCCN 2020043773 (ebook) | ISBN 9780815357131 (hbk) | ISBN 9781351123983 (ebk)
Subjects: LCSH: Aluminum in automobiles. | Lightweight construction.
Classification: LCC TL240.5.A48 E24 2021 (print) | LCC TL240.5.A48 (ebook) | DDC 629.2/32--dc23
LC record available at https://lccn.loc.gov/2020043772
LC ebook record available at https://lccn.loc.gov/2020043773

ISBN: 978-0-8153-5713-1 (hbk)
ISBN: 978-0-3676-2363-0 (pbk)
ISBN: 978-1-3511-2398-3 (ebk)

Typeset in Palatino
by KnowledgeWorks Global Ltd.

Contents

Preface

Automotive lightweighting and joining methodologies has been a subject of interest in the mobility and many other industries for many years, and, recently, has become a more important topic. The author's motivation to write this book, "Primer on Automotive Lightweighting Technologies: Design, Manufacturing, and Joining Methodologies with Aluminum" emerged due to his developing short-term courses, organizing one- or two-week workshops, and delivering invited and keynote talks on this and other similar topics both nationally and internationally. The first five chapters of this book cover the basics of lightweighting with aluminum including design considerations, grades and properties of aluminum, manufacturing methods, and car body structures. The material presented in each chapter is developed by the author based on his class notes, and from a collection of articles available in the literature (referenced at the end of each chapter). Although this book emphasizes aluminum as the choice of material for lightweighting, steel and its alloys are still predominantly used for chassis and other major load-bearing members in mobility industries. In addition, plastics, composites, and magnesium are being used in mobility industries. Industrial applications of these materials are also briefly covered in this book. Joining of similar and dissimilar materials is a current topic of industrial and academic research. Therefore, the last chapter of this book has been dedicated to joining methodologies.

The purpose of this book is to expand the associate, undergraduate, and graduate engineering curriculum with a unique and ever-growing topic on "Design with Aluminum and Other Materials for Automotive Lightweighting Technologies". This book is meant to provide a development of thinking skills needed for undergraduate students or practicing technicians. The reader is assumed to have the entry-level knowledge of mechanical properties of materials, stress and strength considerations, and manufacturing processes. The material presented in this book is very basic and originally developed for 2-year engineering students; however, more rigor in terms of mathematical modeling and simulation using state-of-the-art CAE tools needs to be added to teach the course at an advanced level. The author is constantly working on updating the material by engaging the undergraduate and graduate students who do thesis as a part of their course requirements.

As mentioned before, the author presented papers, delivered invited and keynote lectures, and organized workshops based on the related topic titled "Lightweighting Technologies for Automotive Applications" at several domestic and international technical meetings. The purpose of these presentations was to show the need for using lightweight materials in the mobility industry and to discuss the author's experience in developing and teaching courses on lightweighting, including gauging student success in the courses. This manuscript is the first attempt by the author to assimilate the experiences of developing and delivering this course. It is the intent of the author to add more advanced material of current interest in the future editions of this book.

Summary, in the form of Introduction, is presented at the beginning of each chapter of this book. The author requests feedback from the readers to improve the rigor and content presented in the book. Their suggestions will be duly acknowledged in the future editions of this book, if any.

Acknowledgments

The following persons/organizations helped me in the development of this book. Their help is sincerely acknowledged.

1. The Center for Advanced Automotive Technology (CAAT) at Macomb Community College (MCC), Warren, MI: the author developed and delivered the original version of a portion of this course for MCC's Engineering Technology Associate Degree students. This part of the course material was based on an automotive lightweighting theme developed by Professor Echempati and his colleagues at Kettering in three separate seed-funding contracts between CAAT and Kettering University. CAAT received funding from the National Science Foundation under Grant No. 1400593.

2. Prof. Craig Hoff (Dean of College of Engineering) and Prof. Yaomin Dong, both of Mechanical Department at Kettering University, Flint, MI: for helping me in reviewing parts of the original lecture slides of the course material, "Design with Aluminum" that the author developed and delivered at MCC in 2016.

3. Kettering University: for providing me with unique opportunities both on and off campus to develop my applied research skills by continuously providing me with the support whenever I requested or needed. My job functions provided me numerous opportunities to visit or contact our industry sponsors, partner schools in many countries, and to deliver invited lectures and technical presentations in several professional conferences, etc., that motivated me to write this book.

4. Mr. Daniel Scheda and Mr. Rishab Gupta, both graduate students and Teaching Assistants (TA) of mechanical engineering at Kettering University: they worked under my guidance and expanded the material presented in the book based on the lecture slides and other references provided by the author. Daniel took this course as one of his ME technical electives.

5. Mr. Patrik Ragnarsson and his colleagues at European Aluminium: for allowing me to use lots of material from their 2015 Aluminium Automotive Manual.

6. CRC Press publishers have been patient for a long time for me to prepare this manuscript. This manuscript will be refined and updated at a future time with more theory, analysis, and current information on lightweighting in mobility industries.

7. Finally, I thank my wife, Pankaja Echempati for her immense love and friendship towards me, and my two loving daughters, Sharwari E. Puskala and Aparna E. Bankston: they supported me in various capacities throughout my academic and research career. I dedicate this book to my parents and my other family members.

Author Biography

 **Dr. Raghu Echempati**, PI, is a full professor of Mechanical Engineering at Kettering University (formerly, General Motors Institute of Technology). He earned his Master's and Ph.D. in Mechanical Engineering from the Indian Institute of Technology (India) and another Master's in Engineering Management from Kettering University. He is a registered professional engineer (P.E.). He was an active member of SME and a certified manufacturing engineer also granted by SME. He is an active member and a fellow of ASME, active member and McFarland awardee of SAE, and an active member of ASEE. He participated in several study abroad programs in Germany and taught design and finite element analysis courses for several semesters in Germany and in India. He secured Fulbright award twice to teach in India and in Thailand.

Since 2005, he has been one of the organizers of Body Design and Engineering session of SAE world congress. He is a panel member to review proposals submitted to NSF, Fulbright and Gilman Foundation (for study abroad). He has published over 135 journal and conference papers and supervised more than 180 undergraduate and graduate student theses in the areas of design, finite element analysis and manufacturing. His academic teaching, applied research, and consulting experience spans over 30 years. He taught several core and elective courses in the mechanics, design, vibrations, and manufacture areas, including sheet metal-forming course that he developed and taught for many years at Kettering University. He published a few papers with his students on formability of aluminum sheet metal parts.

Dr. Echempati was a Bosch Professor and workded on a project related to the causes of injuries due to airbag systems. He later worked at GM and other industries as a faculty intern to understand the best practices followed. He brought some of those experiences back to share with the colleagues and students of his calsses. At GM, he worked in the die design and stamping operations unit to understand the real-life problems in bulk deformation and sheet metal-forming processes, primarily using steel and, to some extent, aluminum, formed conventionally and by hydroforming operations. He also worked at Global Engine Manufacturing Alliance (GEMA), now known as Dundee Engine Plant, which is now wholly owned by Chrysler Group LLC, as a faculty intern to understand the error and mistake proofing (EMP) operations used in the assembly line of small engines.

1

Engineering Design with Aluminum

1.1 Introduction

Engineering is purposeful and practical ingenuity employed in the service of mankind. *Design* is an iterative decision-making process of creation and optimization to completely and unambiguously fulfill some human need. Design is a sociotechnical activity that is open-ended; not a subject or coursework. Design projects can be tackled in many different ways and can vary depending on group dynamics, project goals, and existing content knowledge. The design task requires motivation and persistence that are shared by the individuals in the design team.

The *engineering design process* is the application of the design process in the field of engineering that enables management policies to be realized as products; the core of any manufacturing enterprise. The Accreditation Board for Engineering and Technology (ABET) defines engineering design as the process of devising a system, component, or process to meet desired needs. Any engineering design project has the primary objective to fulfill some identified human need. However, an important consideration is the regard for the resource conservation and environmental impact. The engineering design process is often iterative where basic sciences, mathematics, and engineering sciences are applied to convert resources to meet the stated objective. The stated objective is one of the fundamental elements of the design process along with constraints, synthesis, analysis, construction, testing, and evaluation. Case studies in engineering design play a vital role in learning the design process from socioeconomic and safety perspectives. In addition to case studies, simulation of the design stimulates engineers to model more complex but realistic environments. Good facilities provide incentives for creative thinking, team activity, decision making, optimization, and presentation using engineering communication skills. The design process is not restricted to the engineering discipline; however, engineering design requires the realization of engineering products and thus requires an engineering background. A common flowchart for the design process can be found in the literature, such as given in reference [1] and it consists of identifying and defining the problem, doing a literature search to see if the problem is already addressed all the way to develop conceptual designs, performing analysis, building the device, and testing.

1.2 Design Methodology

Design methodology generally refers to the development of a system or method for a unique situation. A realistic design methodology should define global objectives and choices in a concrete way. For example, the "waterfall model" identifies specific phases such as design, implementation, realization, testing, production, and maintenance. In addition to traditional methods (such as the waterfall method), engineering design methodology also uses other methods, such as statistical and artificial intelligence methods, to design products and ideas. References [1 – 4] discuss more details about this topic.

The engineering design process has four distinct phases:

1. Clarify the problem
2. Develop concepts
3. Embody design
4. Detailed design

1.2.1 Phase 1: Clarify the Problem

In order to clarify the problem, research is needed on the topics of interest relevant to the design problem as well as the customer/human needs. A common customer need process flowchart can be found in reference [3] and it mainly consists of data collection and analysis.

The first step in identifying customer needs is to define the scope. Defining the scope involves a description of the product being designed, the business goals (e.g., a product introduced in the summer of 2020, 50% gross margin, etc.), the primary and secondary markets, and the stakeholders (creditors, directors, employees, owners, suppliers, unions, etc.). Stakeholders are individuals who have an interest or concern in an organization.

The next step in the process is to gather data about the problem that the design team is trying to solve. The data can be acquired from the following:

- Information from customers or customer interviews
- Research and development (R&D) department
- Focus groups
- Competitors

- Ethnographic discovery (searching for user patterns and habits)
- Trade shows
- Employees or salespeople

Once the relevant information is gathered from the sources mentioned above, interpreting the data is the next step. Oftentimes interpreting the data can be difficult; however, organizing the gathered data into need statements is a useful tactic. The objective of a need statement table is to use the gathered data to determine what attributes or design considerations should be taken into account for the new product. See an example of a need statement in Table 1.1 shown below. https://slideplayer.com/slide/10129906/

The next step is to organize the needs identified after interpreting the data the team has gathered. Two common methods of organizing the needs are by using the Hierarchical Method or the Kano Method. The hierarchical method organizes the new product needs into primary needs, secondary needs, and tertiary needs. It is up to the team to decide which identified needs fall into what category. The Kano method, shown in Figure 1.1, classifies needs into dissatisfiers, satisfiers, and delighters. *Dissatisfiers* are product characteristics that are basic or expected and are usually taken for granted by customers. *Satisfiers* are product characteristics that customers want in their products and are linear. In this case, linear means the more that is provided,

TABLE 1.1

Need Statement (Adapted from Reference [3])

Guideline	Customer Statement	WRONG Need Statement	CORRECT Need Statement
Start with "What?", rather than "How?"	"Why don't they put a clamp at the end of the outlet hose?"	The outlet hose has a clamp to connect to a water pipe	The water filter easily transfers water into a variety of different containers
Specificity	"Oftentimes I drop the water filter on the floor"	The water filter is robust and rugged	The water filter operates normally after a few accidental droppings
Be positive, not negative	"The water filter is difficult to hold"	The water filter is not difficult to hold	The water filter is made for easy handling
Product attribute	"I need to attach a virus filter to the water filter"	A virus filter can be attached to the water filter	The water filter accommodates a virus filter
Avoid using "must" and "should"	"The water should taste good"	The water filter should deliver good tasting water	The water filter delivers good tasting water

FIGURE 1.1
Kano model [5].

the happier the customer is. Lastly, *delighters* are product characteristics that are attractive or exciting. Delighters pleasantly surprise customers upon initial inspection.

The next step in the process is establishing importance. There are a number of ways of establishing importance. The *Point Direct Rating* involves giving a numerical rating to each identified need, typically on a scale of 1–10. Some thoughts when giving numerical values could be "How important is this feature?" or "Is the feature desirable, neutral, or undesirable?" For this method, the higher the number, the more important the need is. Keep in mind that the frequency of mentioning a need is not always a good measure of the importance of a need.

The last step is reflecting on the process. This process is aimed to capture what the needs are and not how to satisfy the needs. Some of the takeaways could be the experiences of meeting customers in the user environment, interviews are more efficient than focus groups, and the process of collecting visual, verbal and textual data. This is the step that will transition the design process from defining the scope into developing concepts.

1.2.2 Phase 2: Develop Concepts

Developing concepts involves brainstorming different ideas to solve the problem, forming a matrix for a morphological analysis, functional decomposition, generating concept variants, and then evaluating the concepts against each other.

Brainstorming is the first step in developing concepts. One of the best ways to generate good ideas is to get lots of ideas and weed out the ideas that are "bad". Going for quantity and not quality in an atmosphere with no criticism is the goal of any brainstorming session. It is a common misconception that the experts in the subject of interest are the only people who should brainstorm ideas. However, the purpose of brainstorming is to get a wide variety of ideas, so if the only people in the room will be experts, only a narrow scope of ideas will be generated. In other words, invite different types of people from a variety of backgrounds. Setting a time limit is a good way to cut the brainstorm session off before the brainstormers get stale. It may take a few brainstorming sessions before the team is comfortable with the number of ideas before filtering them to find the keepers.

Once the team has generated a sufficient number of ideas to solve the problem, the team must identify the functions that the product or system they are designing needs to perform (i.e., what the product or system does). These are called functional elements. The next step is to figure out which functions are the most and least important. This is done by functional decomposition. Functional decomposition is typically done using a top-down process where the overall function is realized and then broken up into subfunctions. During this step, the inputs and outputs are also defined for each function. An example of the functional decomposition for a coffee maker is shown in Figure 1.2.

As seen in Figure 1.2, level 0 is where you start and it will only contain one block. This one block should be the overall function of your product or system. Level 1 is typically referred to as the main system architecture. This level is where your supporting subfunctions will be. Each subfunction

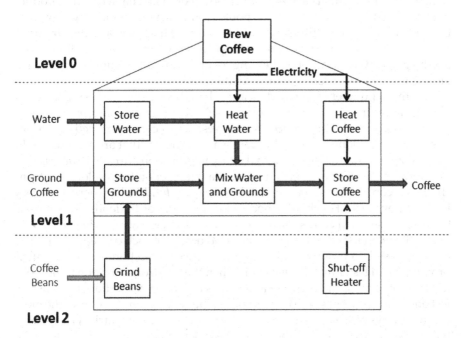

FIGURE 1.2
Example of the functional decomposition for a coffee maker.

TABLE 1.2

Sample Morphological Matrix for a Coffee Maker

	Function		Forms/Solutions to Satisfy Functions				
Brew coffee	Store coffee	S11	S12	......	......	......	S1m
	Mix coffee and water	Filter	Osmosis	Dissolve	Ionize	......	Stir
	Heat coffee	S31	......	......	......	......	S3m
	Heat water	S41	......	......	......	......	S4m
	Store water	Si1	......	......	......	......	Sim
	Store grounds	Sn1	Sn2	......	......	......	Snm

is connected to the flow of the system. This is the level where inputs and outputs are shown and describes the operation (how functions interact with each other). Lastly, level 2 is what describes the auxiliary functions of the system. These are the functions that would be nice to have, but are not absolutely necessary for the system to operate. These auxiliary functions would be classified as delighters from the Kano model.

The team now has a better idea of what the system or product is; a morphological matrix should be made to generate concept variants. A *morphological matrix* is a structured representation technique (Table 1.2) that lists down all the main functions of the product or system and the different physical components through which each of the functions could be enabled in the product or system. The purpose of the morphological matrix is to identify combinations of solutions to fulfill the overall function. This gives a wide variety of potential concept variants of the product to serve the same overall purpose for the product. In addition to helping identify possible solutions, it can also aid the team in discovering which solutions are infeasible.

Now that a number of different concepts have been generated using the morphological matrix, it's on to the concept selection process. *Concept selection* is the process of evaluating concepts with respect to customer needs. Concept selection can be done in a few different ways. Concepts that are turned over to the customer, client, or some external entity for selection are called *external decision*. *Product champion* is where an influential member of the development team chooses a concept based on personal preference. *Intuition* is where a concept is chosen by "feel". For this method, explicit criteria or tradeoffs are not used because the selected concept just "feels" better than the others. *Pros and cons* is where the team lists the strengths and weaknesses of each concept and makes a choice based on group opinion. Another concept selection method is called prototype and test. *Prototype and test* is when the team/organization builds and tests prototypes for each concept and makes the concept selection based on the test data. This concept selection method can be very expensive. One of the most common ways of selecting a concept is by using a decision matrix. When using the *decision matrix* method, the team rates each concept against prespecified selection criteria. This criterion can be weighted depending on importance.

The first three methods mentioned (i.e., external decision, product champion, and intuition) are not rigorous methods of concept selection and often do not produce the "best" products. Using a more structured approach like the last three methods mentioned (pros and cons, prototype and test, and decision matrix) is a more precise way of concept selection. A structured approach also encourages decision making based on objective criteria and minimizes the likelihood that arbitrary or personal factor influences the concept. Lastly, a structured approach more readily allows for documentation. This documentation results in the record of the rationale behind concept decisions and allows for new team members to quickly understand the project thus far.

Another common method for concept selection is called Pugh's method. *Pugh's method* is a structured concept selection process that rates and ranks concepts to develop the most suitable solution. Oftentimes some reference concept is used. This might be an existing concept (e.g., a competitor's product) or some idea the team decides to use as a baseline (e.g., the company's current product). The outline for Pugh's method is presented below.

- Prepare the decision matrix
  - Criteria
  - Reference concept
  - Weightings
- Rate concepts
  - Scale (+, −, 0) or (1–5)
  - Compare to reference concept
- Rank concepts
  - Sum weighted scores
- Combine and improve
  - Remove bad features
  - Combine good qualities
- Select the best concept
  - May be more than one "best" concept
  - Beware of average concepts

1.2.3 Phase 3: Embody Design

After the concept selection is finished, the next step in the design process is embody design. Within the embody design phase, there are three distinct phases:

1. Product architecture
2. Configuration design
3. Parametric design

Product architecture is the arrangement or mapping of functional elements to physical components. Specification of the interfaces among physical components is also part of product architecture. Coupled interfaces are interfaces that if one component requires a change, another change is required to a second component for the overall product to work. If the interfaces are not coupled, they are called decoupled. Depending on how functions are mapped to components (product form) and what type of interfaces exist between the components, product architecture can be classified into different categories: modular and integral. Modular architecture is where there is one physical component per function (decoupled interfaces). Integral architecture is when there are multiple functions per physical component (coupled interfaces).

As discussed in reference [3], there are three types of modularity architecture: slot, bus, and sectional [3]. In slot architecture, each module has a different interface with the overall system. This process does not allow for the various components to be interchanged. This type of modularity is good for the product assembly process since there is one specific location for the components to go. In bus architecture, there is a common bus to which modules connect through a similar interface. This type of architecture can be challenging for product assembly because the modules can connect to every available location even though that could be the incorrect bus for that component. In sectional architecture, all interfaces are the same type, but there is no single element to which the modules can attach (e.g., LEGOS, sectional living room furniture). Using sectional architecture, the product assembly is built up by connecting the modules to each other via identical interfaces.

Configuration design is related to the shape, general dimensions, and orientation of components or physical layout. This is the stage where the design shape takes form. The purpose of this stage is to effectively and efficiently use the allowable design space. This step is particularly important if one of the design constraints is overall dimensions. These initial sketches need not be precise; however, they should show the maximum dimensions of the product, clearance between subsystems, installation paths, and general component arrangement.

Parametric design is the final stage in the embody design phase. In the parametric design stage, exact dimensions and tolerances are applied. In this stage, each component is designed in a Computer-Aided Design (CAD) program. Here is where standard components (screws, fasteners, etc.) are identified. Once the components are drawn, they can be assembled into their subsystems and the final assembly.

1.2.4 Detailed Design

The final phase of the design process is the detailed design phase. The detailed design consists of models and prototypes, detailed engineering drawings, production prototypes, testing, and design review. In order to create a prototype, the first step is to decide which components will be fabricated and which components will be purchased. When the make or buy decisions have been worked out, detailed engineering drawings are needed for the parts to be fabricated. Using a CAD program, such as Autodesk Inventor, Solidworks, or Creo Parametric, the parts can be solid modeled and used to fabricate the components. Once the components are manufactured (through machining, casting, etc., described in Chapter 4), prototypes can be assembled and tested. The final part of the detailed design phase is documentation. This documentation usually comes in the form of a final design report. This report may include the detailed engineering drawings, the process by which the product was designed, the bill of materials (BOM), and the final cost estimates.

1.2.5 Scientific Design, Engineering Design, and Engineering Science

Scientists and engineers sometimes have different objectives and so they follow different design processes for their work. In real-life situations, the distinction between these two methods is not very clear. However, engineers tend to provide usable solutions quickly using prototypes rather than extensive testing of their designs.

The basic principle of a scientific method starts with question or inquiry. This is followed by hypothesis or conjecture. After this, we predict what the expected results might be. This is possible due to previous experiences of success or failure. After this, we conduct experiments or testing to assess if our predictions are going as thought. The final step is to conduct analysis of the results to understand the future course of action which may include redoing experiments or making sure if the initial question is correctly understood. The scientific method can be iterative depending on the case in hand. Table 1.3 shows these steps in a serial manner.

The engineering design process follows similar steps as scientific method; however, it is mostly iterative. Basic steps of a typical engineering process using Lucidchart [6] are shown in the flowchart below (Figure 1.3).

There is a distinct difference between engineering science and engineering design as shown in Table 1.3 and Figure 1.3. Engineering science is a

TABLE 1.3

Typical Steps of a Scientific Design Method

Question or inquiry
Hypothesis or conjecture
Prediction
Experiment or testing
Analysis of results
Future course of action

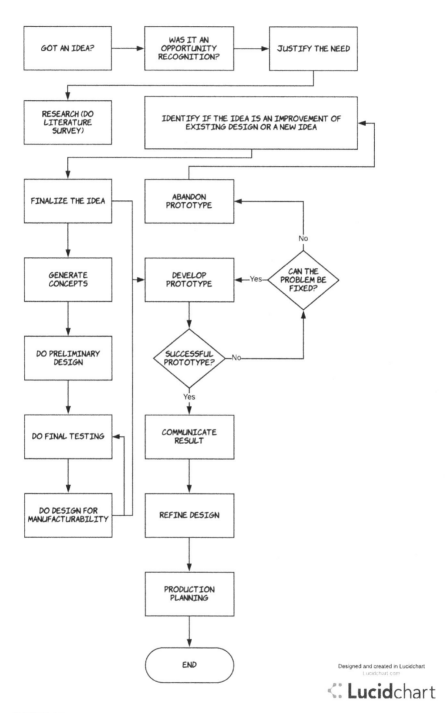

FIGURE 1.3
Typical engineering design processes [6].

closed-ended problem with a unique specialized solution that uses specialized knowledge whereas engineering design is a much more open problem that has no identifiable end with many solutions and requires knowledge from a variety of different fields. In order to be a good engineering designer, one must have a solid background of engineering science. However, a designer must be prepared for ambiguous requirements or specification and perhaps face uncertainties along the way to complete a design project.

1.3 Design for Manufacture and Assembly

Design for manufacturing (DFM) is a method of design for ease of manufacturing of the collection of parts that will form the product after assembly. DFM is a tool that deals with selecting cost-effective material and process(es) to be used in production. DFM is an activity that should be performed in the early stages of engineering design to cut down on cost and manufacturing time.

Design for assembly (DFA) is a method of design of the product for ease of assembly. DFA attempts to devise method(s) to optimize parts or assembly in a more efficient manner to speed the assembly process up or make the assembly easier. In addition, DFA attempts to reduce the number of parts and promotes designing automatic assembly cells and assembly operations. DFA occurs in the automotive industry, particularly in engine assembly. Engine assemblies contain thousands of parts. Making the assembly process more efficient and easier makes the engine assembly process more cost-effective. Figure 1.4 shows a block diagram of an example of the integrated cost-effective model.

1.4 Design for Functional Performance

Design for functional performance (DFFP) is a method of design that pays special attention to safety and societal requirements. In addition, packaging aspects play an important role in lightweight automotive design (i.e., using aluminum). DFFP takes into account material properties, stiffness, crash performance, fatigue, and corrosion issues.

In the case of designing with aluminum for automotive structures and components, knowledge and experience in a variety of areas is desired in order for optimal and predictable performance during service. Some of the areas of interest include:

- Structural stiffness
- Stability and fatigue behavior

- Crash behavior
- Corrosion performance of aluminum alloys
- Protection of fasteners against failure

1.4.1 Galvanic Corrosion

This subtopic provides an overview of galvanic corrosion, its causes, and proven corrosion prevention strategies from initial design to road usage [7-9]. *Galvanic corrosion* is an electrochemical action of two dissimilar metals in the presence of an electrolyte and electron conduction path. Essentially, when two different metals are in contact, the more noble metal (cathode) decreases its corrosion potential at the expense of the more active metal (anode). The corrosion of one metal is decreased while that of the neighboring one is accelerated. A common source of corrosion for automobiles is water or road salt acting as a conducting electrolyte for the potential change.

PPG Industries Inc. reports that every year, businesses in the United States will spend over \$270 billion repairing damage caused by galvanic corrosion. Of these \$270 billion, \$50 billion is in the transportation business alone. Fighting corrosion effectively can be a major source of savings for companies that depend on transportation. Therefore, this makes corrosion performance of aluminum alloys in automotive structures extremely important.

There are a number of factors involved in the corrosion of a bimetallic couple. Reversible electrode potentials can lend themselves to having a greater potential for corrosion. Corrosion potential is referred to as a metal anodic index. The farther apart the two metals are, the stronger is the rate of corrosion. This is important for metal-to-metal contact, especially when the two metals are dissimilar. Some examples that are commonly found in the auto industry are different metals adjoined on hinges, door frames, mounting brackets, and fasteners (bolts, rivets, etc.). Geometric factors, such as surface area, shape, and orientation, can have an effect on how badly and where on the part the corrosion exits. This can be a particularly important factor if there already exists some predisposed failure site in the part. The failure time can be significantly increased with the presence of corrosion. Metallurgical factors, such as alloying and heat treatment, affect how a part resists corrosive effects. Along with the metallurgical properties, the surface conditions of the part are important to corrosive behavior. Films and corrosive resisting coatings are often used to resist the effects of corrosion. Another important factor is the type of environment the part is in. Conditions, such as cyclic moisture and solar radiation, strongly affect how a part resists corrosion. Electrolytic properties also factor into a part's ability to resist corrosion. The ionic species, pH, conductivity, temperature, volume, and flow rate are all properties of the electrolyte that can dictate how a part corrodes. An electrolyte transfers

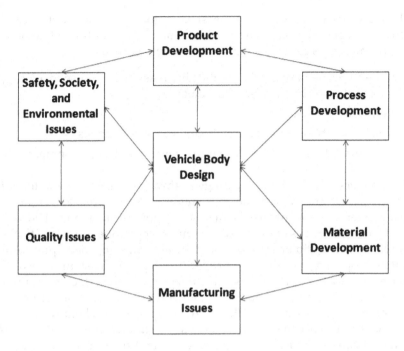

FIGURE 1.4
Integrated cost-effective model.

corrosion from the stronger metal (cathode) to the weaker metal (anode) described by Kirchhoff's Second Law (Equation 1.1).

$$E_c - E_a = I * R_e + I * R_m \qquad (1.1)$$

Where R_e is the resistance of the electrolyte, R_m is the resistance of the metals, E_c is the electric potential of the cathode, and E_a is the electric potential of the anode.

It is easy to find galvanic corrosion in the transportation industry. Critical areas, such as welds, seams, hinges, and fasteners, are the most common. These are common areas for corrosion due to the potential for moisture to linger on there. Some common conductors that accelerate the corrosion process include sodium chloride, calcium chloride, and magnesium chloride. Automakers focus protection efforts on the underbody since these accelerators come in contact with the underbody most often (like the use of road salt) and because the electrolytes also transfer through humidity in addition to direct contact. The corrosion performance of automotive aluminum alloys, such as 319, 356, 380 (for casted parts), and 6061 (for forged parts), is considered acceptable for the expected lifetime of the current automobiles.

There are several preventative measures automakers carry out to combat corrosion. Depending on the application, one or more of the following preventative measures could be employed to reduce corrosion.

1. Avoid the use of dissimilar metals in an assembly or use insulating material between the dissimilar metals
2. Impermeable coatings
3. Avoid situations with large cathodes and small anodes
4. Add inhibitors to reduce the aggressiveness of the environment

Automakers commonly use coatings to thwart corrosion. For autobodies, the body panels are painted and thus automakers use a primer (Figure 1.5). These primers are not only used as a base coat for excellent topcoat adhesion, but also to fight corrosion and resist paint chipping due to rocks, etc. Some examples of these primers include alkyd primer (alkyd topcoat), epoxy primer (urethane topcoat), and zinc-rich primer (urethane midcoat and topcoat). In order to select which primer best suits the customer needs, climate, humidity, exposure time to precipitation, production ease, and cost-performance tradeoff must be considered. Alkyds are cheaper, zinc-rich epoxies provide superior performance (most widely used in automotive industries), and polyurethane primers are used for fast cure speed and smooth appearance.

1.4.2 Fastener Protection, Nonconductive Barrier Material, and Mixed Material Designs

Crevice corrosion takes place at the overlapping of joined parts, welding zones, and under the welding deposits (slag and precipitates). Figure 1.6

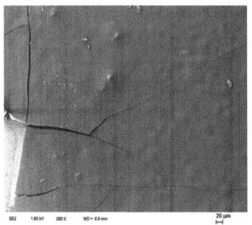

FIGURE 1.5
Cracks of silicone coating (*left*). Paint blistering around hinge (*right*).

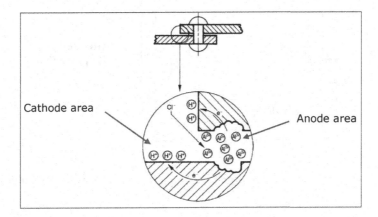

FIGURE 1.6
Crevice corrosion of joined metal parts at low pH value.

shows an example of crevice corrosion. Once the corrosive electrolyte pen- etrates between the joined parts, the corrosion process starts at the anode area. In order to prevent the electrolyte from penetrating between the joined parts, nonconductive barriers are used to combat corrosion. There are two types of barriers: physical barriers and sacrificial barriers (Figure 1.7). Some

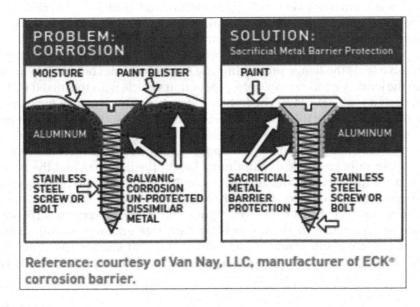

FIGURE 1.7
Use of nonconductive sacrificial barrier materials.

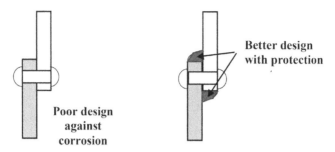

FIGURE 1.8
Example of the best practices for joint protection against corrosion of dissimilar materials.

examples of physical barriers include Mylar tape, synthetic fasteners, nylon washers, and other kinds of nonconductive materials. Sacrificial barriers are the purposeful placement of an anode material directly in contact with the cathode metal creating galvanic corrosion.

Mixed material construction in design exists today and is widely used in the automotive industry. Due to the mixed material designs (usually aluminum and steel), corrosion is very prevalent since steel and aluminum are two dissimilar metals. When bare aluminum joins with bare steel, corrosion can result. So, innovative joining techniques are employed. Figure 1.8 shows an example of the best practices used to combat corrosion. The joints between aluminum and steel are made using adhesives and rivets. This process is known as rivet bonding. In addition to the adhesive barrier, steel panels are galvanized and the aluminum panels are treated with a titanium/zircon coating designed to both hinder oxidation and enhance the adhesive bonding surface. Another benefit of this method is that the cured adhesive seals the flange preventing any electrolyte from being introduced into the joint. A good example of a production vehicle with the mixed material design is a 2004 BMW. The frontend structure from the cowl forward is all aluminum while the rest of the vehicle is steel. The front portion of the lower rails is aluminum and the portion of the lower rails below the floor pan is steel. A rear portion of the upper rails is steel, however, the rest of the upper rails are made of aluminum. Lastly, the floor pan and the lower portion of the A-pillars are steel, both joined to an aluminum cowl panel.

Aluminum, although costlier, is the future direction of many auto manufacturers for its benefits over steel. Aluminum is easily prone to corrosion. The corrosion found in aluminum is typically intergranular. Modifications, such as altering the grain size and shape and introducing elements like Titanium and Zirconium in the grain structure, are industry measures aimed at achieving long life capability.

1.5 Design for Cost Optimization versus Maintenance

Optimum design is defined as the minimum total business impact. In order to achieve an optimum cost versus maintenance, we need to understand whether maintenance is worth doing and when should the maintenance be done. Functional analysis needs to be done first, followed by a critical assessment at different levels of design. No single formula seems well suited to fit a multitude of different industries, or even to different processes, plant types, or departments within the same company. Systematic procedures and analyses are performed to achieve the optimum design based on cost and maintenance. The amount of analysis effort and its payback depend on the importance of arriving at the correct design or strategy desired by an organization. Tools, such as Failure Mode and Effects Analysis (FMEA) and Review of Existing Maintenance (REM), provide simple strategies to verify that there is a valid reason for doing the maintenance job. In addition, these tools are strategies for the cost and interval of maintenance are reasonable in relation to the risk and consequences if the maintenance is not performed. Figure 1.9 below shows the cost/risk impact relative to the time of repairs.

A good example of cost optimization versus maintenance is when it comes to overhauling a pump. If the performance of a pump deteriorates as its impeller becomes foul or worn, and the reduced capacity is having an effect

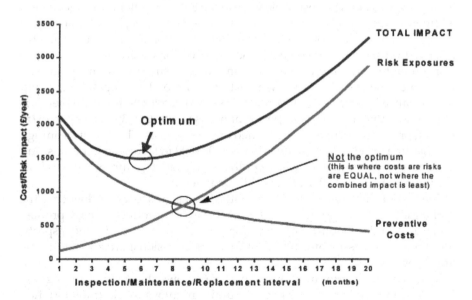

FIGURE 1.9
Factors influencing the maintenance costs in pounds (£) [10].

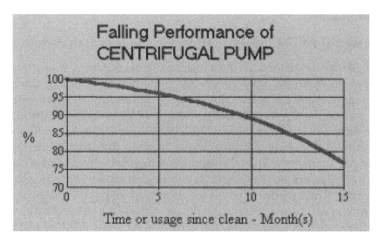

FIGURE 1.10
Pump performance vs. time [10].

upon production or process efficiency, then there must be an optimum time to clean or replace the impeller. To determine the best maintenance strategy, we need to know how the performance decreases with time or usage and the economic effect of the losses (perhaps the pump has to operate for longer to deliver the required volumes or maybe the drive motor draws more electricity to compensate). The cost to clean or replace the impeller is also needed in addition to the operational downtime to complete the repairs. Some of this information may be known if there is some operational experience, but, otherwise, it must be range estimated and explored for sensitivity.

In order to solve the above problem, some assumptions and data estimates must be made. It is known that after 6 months of operation, pump performance is down by 5–10% and is likely to accelerate if left further. It is also known that a 10% loss in performance is worth £10–30 per day in extra energy, production impact, or extended operating costs. The cost of cleaning/ overhauling or replacing the impeller is £600–800 in labor and materials, and 2–3 hours of downtime to swap over to an alternative pump. See the following figures (Figures 1.10–1.12) for additional detail.

The rest of this chapter will deal with the design aspects influencing the economics of aluminum usage in cars. Minimizing production cost, production waste (and its deposition), recycling processes, health and safety aspects, and lifecycle assessment (savings of fuel and emissions) are all factors that affect cost.

Research on aluminum alloys like A 356 shows that overall savings up to 40% are possible with the use of aluminum alloys in the transportation industry. These studies are based on how aluminum is processed (stamping, casting, forging, extrusion, etc.) and tested for suitable mechanical properties

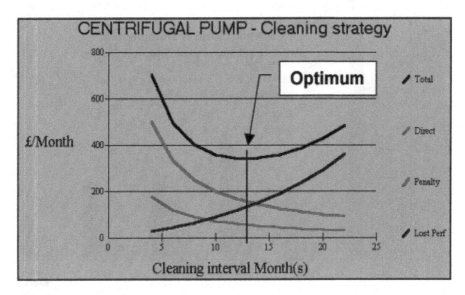

FIGURE 1.11
Optimization cost determination [10].

(strength, fatigue, stiffness, impact, etc.). Usage of aluminum is gaining attention of the original equipment manufacturers (OEMs), although there are some technical barriers in their mass production (see Figure 1.13). https://www.greencarcongress.com/2014/06/20140610-ducker.html

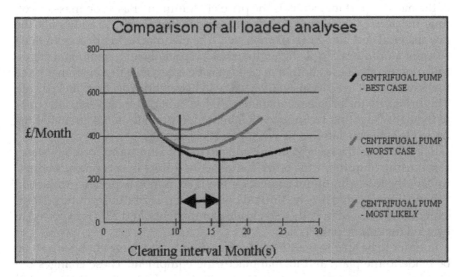

FIGURE 1.12
Sensitivity analysis [10].

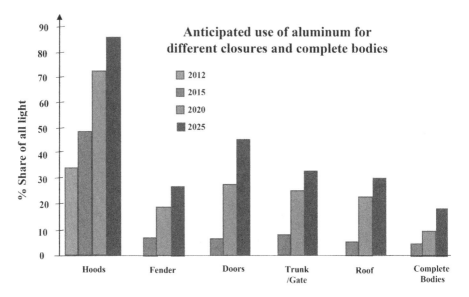

FIGURE 1.13
Previous and anticipated aluminum usage in closures and complete bodies [adapted from 12].

1.5.1 Price of Aluminum Alloy, Weight Savings, and Secondary Cost Savings

When looking to use aluminum as an alternative material for automotive components, there are cost factors associated with the change in material. One of the most important factors is the price of aluminum. Based on mass or volume, the price of aluminum is high compared to that of steel. In addition to the raw material cost being higher than steel, the manufacturing-process-related costs for aluminum can be higher than steel, especially for a traditional manufacturing plant. These additional costs may be acceptable to a customer if the fuel savings that are promised (due to the lightweighting technology) pay off. This brings up the problem of balancing the weight of the car versus the costs of the lightweighting technology. Reference [11] provides a discussion of the tradeoffs between automobile weight and cost savings.

In addition to the weight savings due to use of aluminum for body structures, using aluminum for other components results in secondary savings. In 2015, the total aluminum content for the 17.46 million expected production vehicles increased to nearly 7 billion pounds. Of this 7 billion pounds, 11% is expected to be used in body and closure components, and 33% is expected to be used for engine parts [11].

In order to be able to manufacture aluminum parts, some changes have to be made to the plant and its manufacturing equipment. These changes can be a huge cost for an enterprise. Again, the tradeoffs between the expected part cost and the investment cost to be able to manufacture the parts need to be understood [11]. In addition to this, the question is approximately how long will it take to pay off the investment cost. Literature, such as reference [11], provides the answer that the more parts you manufacture, the cost price per kilogram decreases.

1.5.2 Integration of Multiple Part Designs into a Single Product, Aluminum Extrusions and Their Design Guidelines

Integration of multiple parts in designs reduces the number of fasteners, can make the overall structure stronger, and reduces weight. Enterprises like the Dana Corporation are implementing lightweighting strategies, one of which is design integration. Dana Corporation reduced the number of components in the engine, transmission, and drivetrain assemblies to reduce the weight but kept the same functionality. In general, the value of lightweighting decreases from front to back and from top to bottom in car assemblies. High-pressure die casting is another way companies are able to deliver complete assemblies to reduce overall weight. An example of part integration described above is an aluminum bumper using an extruded part instead of welded sheet metal parts.

The cost of scrap generation and reuse of aluminum parts makes the supply of near-net-shape parts, such as prefabricated extrusions, an attractive choice. Also, the productions of ready-for-assembly parts by the aluminum suppliers also allow them to manage scrap and its recycling more efficiently. Table 1.4 below shows the extruded part selection based on production volume.

Aluminum extrusions allow the production of complex shapes and satisfy the functional requirements, along with providing increased stiffness of the structure, reducing the amount of machining, and facilitating assembly

TABLE 1.4

Selection of Process for Aluminum Extrusions

Production Volume	Types of Extruded Parts
Small (10–1000 units)	Use standard profiles and cross-sections
Low (1,000–10,000 units)	Use of specifically designed ("tailored") profile cross-sections, dedicated extrusion tools, 2D forming in simple tools, simple machining (cutting and drilling), and/or punching
Medium (10,000–100,000 units)	Use of weight-optimized, specifically designed cross-sections, dedicated extrusion and forming tools, standard presses and equipment, punching (instead of machining)
High (> 100,000 units)	Highly optimized extrusion product, dedicated forming tools (bending, hydroforming), punching (very little machining), possibly invested into dedicated machines and equipment

operations, as well as saving cost and reducing weight. Examples of these can be found in [11].

Aluminum extrusion provides a solution for complex geometrical shapes that require little or no fabrication and can do the work of several components joined together while providing a substantial saving in weight at the same time. However, there are variables that come into play with aluminum extrusion: wall thickness and cost-effective production. The wall thickness can be uniform which is easy to produce or variable to increase bending stiffness. The factors which have an effect on the wall thickness are the extrusion force and speed, the choice of aluminum alloy, the shape of the profile, desired surface finish, and tolerance specifications. Cost-effective production is always an important factor and the extrusion should be as production-friendly as possible.

At the design stage, secondary fabrication processes, such as bending, and finishing operations, such as turning, drilling, milling, etc., must be considered for aluminum extrusions and how they affect the overall economics. Concepts related to finishing operations will be discussed in a later chapter; bending will be discussed briefly here.

When planning a profile bending operation, there are many factors to take into account. The type of alloy, temper, and cross-section are important things to consider. The expected tolerances and overall space required for secondary operations need to be taken into account. Surface finish characteristics are also important to keep in mind if they are going to be showing. Lastly, knowing what mechanical strength is required for the design is a critical factor when planning a bend profile.

There are three main types of bending processes: push, rotary, and roll bending. Push bending [12] is when an aluminum extrusion is pressed by a ram into a set of dies to pinch angles into the aluminum extrusions. Rotary bending is where a rotating bending die bends the aluminum extrusion into the desired shape as shown in Figure 1.14 [13]. Finally, roll bending is where three or more rollers bend parts into the desired shape as shown in Figure 1.15 [13]. Roll bending can only be used for parts with relatively large radii and are limited to bending in a single plane. Despite these drawbacks, the bend precision is excellent and the process is very good with symmetric profiles.

1.6 Design for Reliability

What is design for reliability? According to IEEE 610.12.1990, reliability is "the ability of a system or component to perform its required functions under stated conditions for a specific period of time" [5]. Design for reliability (DfR) is an integrated activity in which one should set goals at the beginning of a program and then develop a plan to meet those goals. Product design is

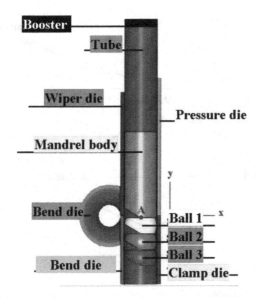

FIGURE 1.14
Schematic of rotary draw bending concept [12].

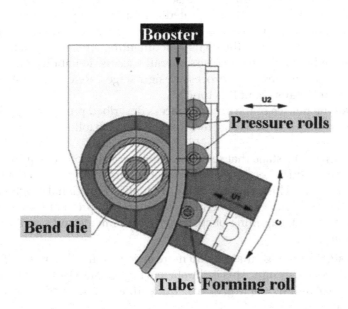

FIGURE 1.15
Schematic of three-roll bending concept [12].

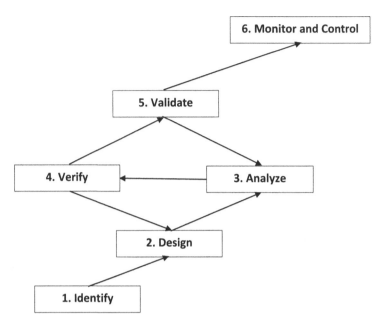

FIGURE 1.16
Design for reliability activities.

not a silo of activities within each discipline, such as mechanical, electrical, chemical, etc., rather, it is an integrated activity encompassing the different aspects of a product or program. The team in charge of DfR should provide metrics and plans to meet the goals. Documenting a reliability plan is a necessary ingredient of an enterprise's overall success. Reliability plans should be included in product and process design stages. References [14] and [15] present more details about this topic.

Just like the phases of the design process described previously, DfR can be broken into stages as well (Figure 1.16). These are as follows:

1. **Identify:** The stage that sets up the goals for reliability requirements of the products, end-user usage, test requirements, and technology limitations. This is where doing a Quality Functional Deployment (QFD) defining requirements, benchmarking, product usage analysis, and understanding of customer requirements and specifications is defined.

2. **Design:** The stage that sets up more specific reliability requirements and risks of design. This is the stage where Design Failure Mode and Effect Analysis (DFMEA), cost tradeoff, and tolerance analysis are done.

3. **Analysis:** This stage requires working with design teams to identify all the potential risks of product usage. During this stage, Finite Element Analysis (FEA), Warranty Data Analysis, and reliability predictions happen.

4. **Verify:** The stage where prototypes are ready for testing for any design modifications. This is the stage where your prototype is tested through evaluation testing, and reliability growth modeling is done.

5. **Validation:** This stage assures the production readiness of the products. This stage gives reliability demonstrations, as well as performs design and process validation.

6. **Control:** The stage that assures the process remains unchanged with minimal variations. Lastly, this stage revalidates manufacturing processes, control charts, and reflection on the processes used for design and implementation.

Another important consideration in designing for reliability is cost. How much is an enterprise willing to spend on reliability? Would you rather pay to service and fix your component or product, or pay to design a more reliable product? This conundrum plagues any company that designs and services its products. Failures in the automotive industry occur frequently. Nothing is more frustrating than being stranded on the side of the road because your car failed you. Making products more reliable makes consumers pleased. The following diagram (Figure 1.17) illustrates reliability versus cost of design.

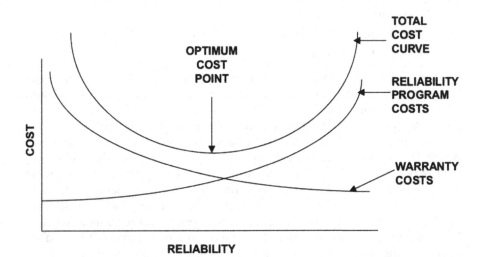

FIGURE 1.17
Reliability vs. design cost.

1.7 Life Cycle Assessment and Sustainability

Sustainable engineering is meeting the needs of the present without compromising the ability of future generations to meet their own needs. What as engineers do we aim to sustain? Some of the common issues of sustainability in design are:

- Environment-friendliness
- Biodegradability
- Ozone-friendliness
- Recyclability

Industry is looking for ways to make their products and processes "green" to make the world more sustainable for the human population. In order to determine if the products and processes are "green", a life cycle analysis (LCA) is performed. LCA is an evaluation of the inputs, outputs, and their potential environmental impacts of a product or process throughout its life cycle. This is done by strategic planning, public policy-making, and marketing eco-friendly solutions. The goal of an LCA is to minimize the inputs and harmful outputs. Common inputs are energy, material, and labor while common outputs include products (electricity, materials, goods, services, etc.), waste, emissions, and coproducts. LCAs start with life cycle inventory analysis (LCI) where the following are defined:

1. What are the function and functional unit?
2. Where are the boundaries?
3. What data is required?
4. What assumptions are made?
5. What are the limitations?

Defining function and functional units is the first part of an LCI. The function is exactly what it sounds like: what the product does. The functional unit gives the function a numerical value. This allows for the comparison between products and a reference point. For example, a pencil's function is to write and a functional unit could be defined as 1 meter of writing per pencil. Data collection is the next important thing in an LCI. When collecting data, experiments are designed and a well-defined scope of the experiments is needed. Once the scope and experiment procedure are defined, the data collection can begin. Some things to keep in mind during the data collection process are consistency, technology, reproducibility, and representativeness. When talking about consistency, does the performed experiment procedure match the designed experiment? As far as the technology is concerned, are the best available processes or technology being used for the experiment? Does the experiment accurately represent the population of interest? Finally, can another agency or person reproduce your data following the same experiment (reproducibility)?

After the LCI is performed, an impact assessment (LCIA) is performed where the environmental, social, and economic effects are determined. The LCIA is done using the data collected from the designed experiment. The most common impact that is studied is the environmental impact. The four main environmental impacts are described in Table 1.5.

The purpose for all this data is so that scientists and engineers can make recommendations on ways to have less impact overall, analyze a particular impact and find a solution, and make a difference on a small or even global scale. References [16] and [17] present more details on this topic. Other useful information for the topics and the subject matter covered in this chapter are given in references [18 to 22].

1.7.1 Scrap Aluminum as Part of Life Cycle Assessment

Aluminum scrap is an important parameter to monitor both for raw material cost as well as Life Cycle Assessment calculations.

It can be seen in Table 1.6 that the scrap value for aluminum is 10 times higher than that of steel. The cost savings for aluminum catch up to steel during the usage phase and surpass steel during the dismantling stage because of aluminum's scrap value. Table 1.7 below gives a summary of the scrap value of aluminum.

TABLE 1.5

Environmental Impact Categories

Environmental Impact	Contributors
Global warming potential	Gases in the atmosphere that absorb and emit radiation
	Trap heat from the sun
	Water vapor, carbon dioxide, methane, ozone, and nitrogen dioxide
Abiotic depletion	Consumption of nonliving resources
	Human toxicity level
	Value that shows harm to humans from chemicals
	Land use
Eutrophication	Increase in chemical nutrients containing nitrogen or phosphorus
	Land or water
	Overgrowth of plants
	Killing organisms at the bottom of the water
Acidification	Pollution from fuels and acid rain
	Low pH

TABLE 1.6

Assumed Data for Life Cycle Assessment Calculations

Parameter	Starting Value	Range
Gas price ($/gal)	2.3	1.84–2.76
Cost of steel ($/kg)	0.9	0.63–1.17
Cost of aluminum ($/kg)	3.3	2.31–4.29
Price of scrap ($/kg)		
Steel	0.09	0.069–0.129
Aluminum	0.93	0.657–1.221
Fuel consumption (mpg)		
Steel BIW	20	
Aluminum BIW	22	
Total vehicle weight (kg)		
Steel BIW	1418	
Aluminum BIW	1155	
Body-in-white weight (kg)		
Steel BIW	371	
Aluminum BIW	193	
Life of the car (years)	14	
Miles driven in year 1	15220	
Lifetime miles driven	174140	
Recycling percentage		
Steel	90	
Aluminum	90	

TABLE 1.7

Summary of the Scrap Value of Aluminum

Item	Price ($/lb)
356 auto wheels	0.700
5052 alum clip	0.680
6061 alum extrusion	0.639
6063 extrusion	0.690
Al/Cu radiator	1.147
Alum breakage 50%	0.193
Alum litho sheets	0.639
Alum old sheet	0.538
Alum rad with iron	0.305
Alum radiator	0.487
Alum transformer	0.102
Alum transmission	0.183
Alum turning	0.518
Mix alum casting	0.558

References

1. The Mechanical Design Process by D.G. Ullman, 4th Edition, McGraw-Hill, 2009.
2. Engineering Design Process by Y.Y. Haik and T. Shahin, 2nd Edition, Cengage Learning, 2011.
3. Product Design and Development by K.T. Ulrich and S.D. Eppinger, 6th Edition, McGraw-Hill, 2015.
4. Engineering Design - Systematic Approach by G. Pahl, W. Beitz, J. Feldhusen and K.-H. Grote, 2nd edition, Springer, 1996.
5. Kano Model. Available at: https://en.wikipedia.org/wiki/Kano_model.
6. https://www.lucidchart.com/documents/editNew/ e7edceca-7aba-4350-aba7-29bbf1599cd9#.
7. "Uhlig's Corrosion Handbook." *Google Books*, books.google.com/books? id=-kK1I fAMJUsC&pg=PA123&lpg=PA123&dq=kirchhoff's law for corrosion& source=bl &ots=g7t9gW9_LZ&sig=ACfU3U2l1ZKL0UUHPMKizo0foh0CXDcCTA&hl=en &sa=X&ved=2ahUKEwjM1fj00vrfAhVmzIMKHVzaCcsQ6AEwFXoECAUQAQ# v=onepage&q=kirchhoff's law for corrosion =false. pg135.
8. Corrosion Engineering: Principles and Practice by P.R. Roberge, McGraw-Hill, 2008.
9. Corrosion of Aluminum and Aluminum Alloys (#06787G), Editor(s): J.R. Davis, ASM International, 1999
10. Cost/Risk Optimisation (C/RO) by John Woodhouse. Article available at: http://www.plant-maintenance.com/articles/Costriskop.pdf, 1999
11. Design for Cost Optimization by The Aluminum Auto Manual, 2011.
12. https://www.cmrp.com/
13. https://commons.wikimedia.org/wiki/File:Three_roll_push_bending_process.jpg.
14. Accendo Reliability Follow. "Design for Reliability (DfR) Seminar." LinkedIn SlideShare, 31 July 2012, www.slideshare.net/fms95032/design-for-reliability-dfr-seminar-13813404.
15. Sustainability Lecture by Cliff Davidson, Syracuse University.
16. Life Cycle Assessment - A Scientific Way to Look at Going Green by M. Lepech, Stanford University.
17. Life Cycle Cost Analysis: Aluminum versus Steel in Passenger Cars, by Ungureanu, et al., TMS, 2007.
18. http://novelis.com/sustainability/manufacturing/
19. Cost-Effectiveness of a Lightweight Design for 2020-2025: An Assessment of a Light-Duty Pickup Truck by C. Caffrey, et al., SAE 2015.
20. The Decisions of Engineering Design by B.L. Marples, Institution of Engineering Designers, 1960, pg 1–16.
21. The Design of Structures of Least Weight by H.L. Cox, Pergamon Press, 1965.
22. Materials, Design and Manufacturing for Lightweight Vehicles by P.K. Mallick, Woodhead Publishing Ltd. (CRC), 1st edition March 31, 2010.

2

Material Considerations

Material selection is a crucial step in the process of design and manufacturing. Different materials have different properties and will perform better in some conditions over others. When designing with aluminum, there are many alloys and tempers to choose from. Economic analysis, cost modeling, and life cycle assessment of the aluminum products can help narrow down which material is "best". Some parameters, such as cost, ease of assembly, production volume, function requirements, ease of operation, and safety, are important considerations when selecting a material. In this chapter, aluminum is compared with steel when being used in automotive applications.

The traditional method of selecting a material is based on stress/strength considerations and based on the applications where they are used. Engineers are presented with the complex problem of choosing from tens of thousands of materials all with different properties and many different manufacturing processes in addition to considering the expected performance of the possible material. No engineer can expect to know more than a small subset of the material properties; therefore, using a material selection procedure can aid the process. Identifying the processing method (casting, machining, welding, forging, etc.) can reduce cost if the material is identified early in the design process. Material cost, material availability, safety, aesthetics, and environmental impact are also important factors to consider. Failure modes are another important factor to consider (over 30 different failure modes in Jack Collins' *Failure of Materials in Mechanical Design*). Based on the expected failure modes for a given component, the following properties of engineering materials are considered.

- Strength (yield, ultimate tensile, shear)
- Ductility
- Modulus of elasticity
- Poisson's ratio
- Hardness
- Creep
- Temperature behavior
- Density
- Anisotropy
- Fatigue strength

- Fracture toughness
- Conductivity

However, obtaining numerical values for the above properties can be challenging but the Ashby procedure can help this.

2.1 Ashby Selection Guidelines

The Ashby methodology is a process used to identify possible material selections for a mechanical application and it has four steps:

1. Translation
2. Screening
3. Ranking
4. Supporting information

Translation is where the design requirements are expressed as constraints and objectives. Screening is where materials that cannot meet the design requirements are eliminated. Ranking assesses each material based on how they fulfill the design requirements in order to find the "best" material. Lastly, supporting information is where the exploration of the pedigrees of top-ranked materials is done.

2.1.1 Translation

As mentioned above, the first step into this process is where the design requirements are identified and expressed as constraints and objectives. The following items are analyzed against the design requirements:

- Function
- Objective
- Constraints
- Free variables

The function identifies what the intended purpose of the component is. A common mistake in this step is to limit the options by specifying implementation within the function. This limits the possibilities that could potentially be "winning" solutions. An objective states what essential conditions must be met, and what manner should the implementation excel. Constraints identify what qualities should be maximized or minimized in addition to differentiating between the bindings as soft constraints. A binding constraint is a constraint that ensures the optimal solution and changes to this constraint change the optimal solution [1]. A soft constraint is a constraint that designers would like to meet but not at the expense of other constraints. Lastly, the designers have to identify the free variables. These free variables can be identified by seeing what qualities of a component are desirable and which qualities can be modified. The following example illustrates the function step:

Ex 2.1: It is desired to select materials for a *light but strong* tie-rod component that is loaded axially.

- *Function:* Support a tension load
- *Objective:* Minimize mass
- *Constraints:* Specified length, and carry a load (F) without failure
- *Free Variables:* Cross-sectional area, and material

The following two equations represent the objective and a constraint respectively [1]:

$$m = \rho * A * L \tag{2.1}$$

$$\frac{F}{A} < \sigma_y \rightarrow A > \frac{F}{\sigma_y} \tag{2.2}$$

By rearranging Equations 2.1 and 2.2, one free variable (the cross-sectional area) can be eliminated as shown in Equation 2.3:

$$m > \frac{F * L * \rho}{\sigma_y} \tag{2.3}$$

With this rearranged equation, the mass can be minimized by minimizing ρ/σ_y or, on the other hand, maximizing σ_y/ρ.

2.1.2 Screening

In the screening step, designers evaluate and assess a large range of materials and eliminate the materials that are insufficient. In order to complete this step, there needs to be an effective way of evaluation and elimination of various material classes and material properties. The main material classes include metals, polymers, glass, ceramics, and composites. Figure 2.1 shows how these different material classes interact with each other to form hybrid materials including metal composites (MC), ceramic composites (CC), polymeric composites (PC), glass-ceramic composites (GCC), and fiber-reinforced glass (FRG).

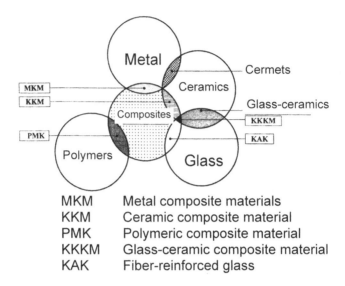

MKM	Metal composite materials
KKM	Ceramic composite material
PMK	Polymeric composite material
KKKM	Glass-ceramic composite material
KAK	Fiber-reinforced glass

FIGURE 2.1
Material classes [adapted from Reference 1].

There are several Ashby charts that rank materials based on various material properties. Screening charts based solely on modulus of elasticity (E) or modulus of elasticity (E) versus density (ρ) are commonly used during the screening process. Using these charts to immediately eliminate materials that are unsuitable for the given application is extremely useful. Figures 2.2 and 2.3 illustrate how each class of materials is related to each other when plotted according to modulus of elasticity versus density. Specifically, in the second figure, the metals and alloys class is highlighted and compares some different metals and alloys within the metals and alloys class.

2.1.3 Ranking

Engineers want to find all the material candidates that satisfy the objective. Returning to Ex 2.1, the objective was to reduce the weight, in this case, minimizing the quantity ρ/σ_y. One common method of ranking, which was discussed in Chapter 1, is to create and use a spreadsheet to screen and rank the materials (see the end of Section 1.1.2). Another method is to utilize the various Ashby charts to rank the materials against each other. Seeing the modulus of elasticity versus density chart (Figure 2.2), it can immediately be seen that metals (and their alloys) and ceramics are the strongest but also the heaviest, where foams are the lightest but very weak by comparison. Additionally, the chart shows that ceramics are somewhere between the foams and metals.

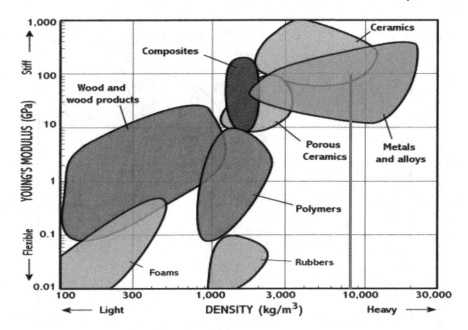

FIGURE 2.2
Ashby's modulus of elasticity vs. density plot [1, 2].

The modulus of elasticity versus density chart is not the only Ashby chart that should be taken into account when selecting a material. Other important Ashby charts are the modulus of elasticity versus cost chart, the strength versus cost chart, the strength versus temperature chart, and the energy versus cost chart (Figure 2.4). The modulus of elasticity versus cost chart shows that metals and alloys are an excellent candidate selection for being stiff as well as moderately cheap.

The strength versus cost chart given in Figure 2.5 shows that composites take a back seat when compared to metals and porous ceramics. This is due to the fact that composites are not nearly as strong or as cost-friendly as compared to metals and porous ceramics.

Temperature can also be an important factor when ranking and selecting a material. Engine components especially undergo various thermal conditions and need to be able to withstand extremely high temperatures during the combustion process. The strength versus temperature chart given in Figure 2.6 indicates that ceramics can withstand extremely high temperatures in addition to being very strong. However, they are closely followed by metals and alloys which can also withstand high temperatures and having high strength.

Overall, it can be seen that metals and alloys have been and still are an excellent choice for a wide variety of loading at a reasonable cost. This is important to automotive components that are subjected to numerous

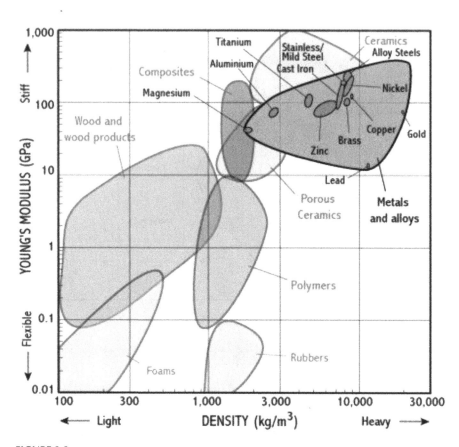

FIGURE 2.3
Details of metals and alloys [1, 2].

different loading environments including axial, bending, and torsional loading scenarios.

Another more advanced method of material selection is called the material index. This method identifies function, constraints, objective(s), and free
variables just like the Ashby method but the material index uses a performance equation to identify and rank materials. If the objective involves a free
variable, the constraint that limits the free variable is identified and used to
eliminate the free variable in question. This gives us the performance equation for a given application. Once the performance equation is derived, the
engineer decides which material properties to maximize performance and
then use the material index for ranking. Table 2.1 shows a few performance
equations for different types of loading. The performance metric for this
table is mass and the objective is to minimize the mass by maximizing the
parameters in the table.

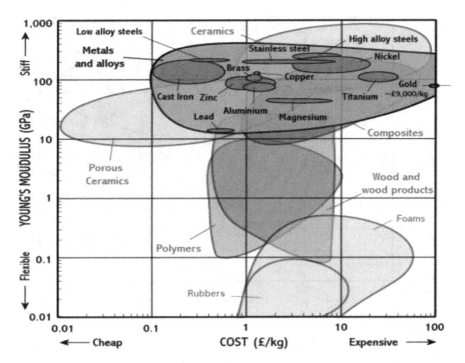

FIGURE 2.4
Modulus of elasticity vs. cost [Ashby chart, Ref 1, 2].

Returning to example 2.1 (axially loaded tie rod component), the objective was to minimize the mass by maximizing $m = {\sigma_y}/{\rho}$, as shown in Equation 2.4 below. This equation can be rewritten as:

$$\sigma_y = m * \rho \qquad (2.4)$$

By taking the logarithm of both sides, we get the following:

$$log(\sigma_y) = log(m) + log(\rho) \qquad (2.5)$$

Equation 2.5 is the equation of a line with a slope of 1 on a log-log scale. This can be plotted on a strength versus density chart (Figure 2.7) to identify the candidate materials. Anything above the line is a candidate that sufficiently minimizes the mass.

For the case of a bending load, stiffness is the consideration and maximizing $m = {E^{0.5}}/{\rho}$ is the driving factor, as shown in Equation 2.6 below. Converting this equation to log space yields:

$$log(E) = 2 * [log(m) + log(\rho)] \qquad (2.6)$$

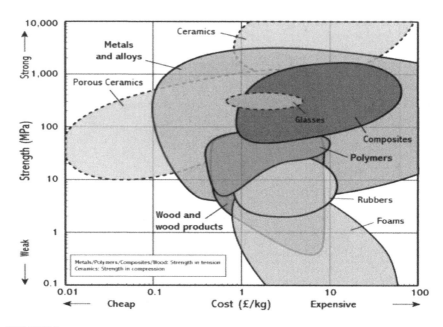

FIGURE 2.5
Ashby's strength vs. cost chart [1, 2].

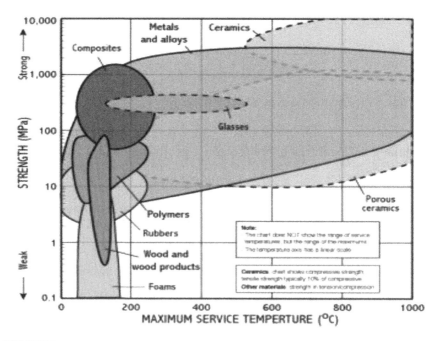

FIGURE 2.6
Ashby's strength vs. temperature chart [1, 2].

TABLE 2.1

Material Indices for a Beam [2]

Loading	Stiffness Limited	Strength Limited
Tension	E/ρ	σ_y/ρ
Bending	$E^{0.5}/\rho$	$\sigma_y{}^{0.666}/\rho$
Torsion	$G^{0.5}/\rho$	$\sigma_y{}^{0.666}/\rho$

↑ *Maximize* ↑

This is a line on a log-log scale that has a slope of 2 and, therefore, anything above the solid line is candidates. This is shown in Figure 2.8.

For design problems that have multiple constraints, the methodology is to solve each constraint possibility individually against the objective. This is done by solving the performance equations and plotting them on the relevant charts to see where the lines intersect and give the class(es) of materials that are suitable candidates. The candidates should be selected based on the most limiting constraint.

2.1.4 Summary of Ashby Procedure

Using the Ashby procedure can be extremely useful when engineers select materials. The procedure has demonstrated that the selection of material affects the design based on geometric specifications, loading requirements,

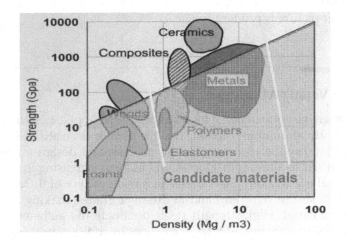

FIGURE 2.7
Optimum selection process for axial loading example [1, 2].

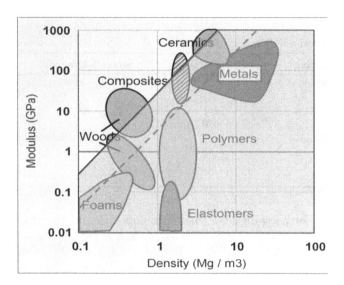

FIGURE 2.8
Optimum selection process for bending load [1, 2].

design constraints, and performance objectives. These effects can be analyzed analytically by using performance equations. It has also shown that keeping the candidate set large will give the "best" results as long as it is feasible to keep the material set large. The Ashby charts can give quick overviews that help engineers and designers to eliminate unsuitable options. Based on the Ashby procedure for material selection, the traditional design process should be modified to include the material selection and process constraints (Figure 2.9).

While the Ashby procedure is fairly robust and will often produce good results, design is holistic where everything matters all at once; so, thinking of the bigger picture is prudent.

2.2 Steel Versus Aluminum

Steel has been the go-to material for many structural applications for over 100 years. Recently, aluminum has come into the picture as an attractive alternative to steel. Since the use cases, design process, and design problems for steel are similar to aluminum, aluminum has been increasing in popularity for structural applications. Steel comes in a wide variety of different forms (rolled stock, sheets, wire, etc.) and in different grades, making it a widely applicable material. High-strength steels dominate the automotive industry today due to increases in total elongation. In addition to those benefits, steel is also more stable against buckling and has higher fatigue strength compared to aluminum. Lastly, steel has very well-known heat treatment

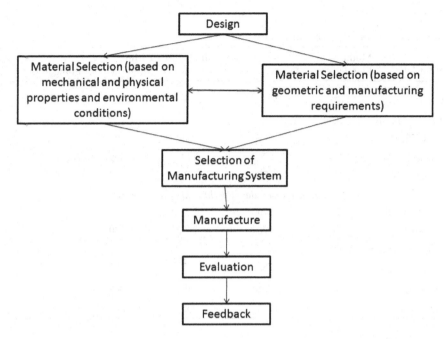

FIGURE 2.9
Alternative design decision model.

property tables that can give an insight into the grain size as well as the solid solution phases (martensite, austenite, etc.).

Strength is not the only material parameter to be considered when considering materials for automotive applications. The density has become an important parameter for lightweighting as well as corrosion resistance properties for reliability. The stiffness and density of aluminum are about one-third that of steel. This makes them an attractive alternative when trying to reduce the weight of automotive structures (Table 2.2). Additionally, aluminum alloys can exceed the strength level of low-strength steels as well as compete with high-strength steels (dominate material in auto industry). These high-strength aluminum alloys are currently being used in aerospace applications. Lastly, on a weight-specific basis, steel and aluminum have similar levels of the modulus of elasticity.

There are numerous parameters that come into consideration when comparing steel and aluminum. Some of these might include:

- Strength-to-weight ratio
- Corrosion resistance
- Thermal conductivity
- Cost

TABLE 2.2

Mechanical Properties of Aluminum and Steel

Material	Modulus of Elasticity	Density	Yield Strength	Ultimate Strength	Elongation
Units	GPa	kg/m^3	MPa	MPa	%
Typical Aluminum Materials					
AA 5754-O	70	2700	110	220	23
AA 6016-T4	70	2700	100	205	27
AA 6111-T4	70	2700	135	275	25
The properties after 2% strain and simulated paint baking treatment.					
AA 6016-T8x	70	2700	220	270	16
AA 6111-T4x	70	2700	260	330	19
Typical Steel Materials					
$FePO_5$	205	7850	155	295	49
IF Steel	205	7850	210	365	36
DP Steel	205	7850	335	500	24

- Workability
- Welding capability
- Thermal properties
- Electrical conductivity
- Strength
- Material availability

When comparing the strength-to-weight ratio of steel versus aluminum, aluminum is typically not as strong as steel but is about one-third the weight of steel. The lightweight nature of aluminum is becoming more attractive to the automotive industry and has been used in aerospace applications for years. Corrosion resistance is another important consideration between aluminum and steel. Aluminum has high oxidation and corrosion resistance due to its passivation layer. When aluminum is oxidized, the surface becomes white and will form pits. However, once this first layer of corrosion forms, it provides a protective layer from further corrosion unlike steel. However, if either steel or aluminum is subjected to extreme acidic or basic environments, the materials will corrode rapidly and can have catastrophic failures. Stainless steel is an alloy of steel that contains chromium which is a noncorrosive material. Stainless steel is not as strong as other forms of steel and it's much more expensive. So, it is sparingly used. Thermal conductivity is another important parameter to consider. Aluminum has a higher thermal conductivity than that of steel. This is one of the reasons aluminum is widely used as radiators and air conditioning units. In addition, having a higher thermal conductivity allows more heat transfer to occur when machining which protects machine tools in the long run. In that same vein, aluminum is more ductile than steel, making it much easier to form, cut, and machine. However, this ductility allows for small dings and damage to occur more easily compared to steel. In the manufacturing process, welding is often an attractive joining method and is much easier to do with steels and can be very difficult to do with aluminum. Aluminum requires special environments for welding. Thermal properties are important for high-temperature applications. Aluminum becomes very soft above 400° C where steels can withstand much higher temperatures. Electrical conductivity is important when the need for passing current is present. Aluminum is a very good conductor of electricity where steel is not as conductive. Due to the electrical conductivity in addition to aluminum's corrosion resistance, research is being done to use aluminum cables for automotive applications. Lastly, cost is a paramount consideration for any enterprise. Steel has been a cheap material for a very long time, but the reduction in aluminum cost has enabled aluminum to compete with steel. Aluminum is even cheaper than some steel alloys, such as stainless steel.

In conclusion, aluminum has numerous benefits that would enable it to be widely used in the automotive industry. However, aluminum does have drawbacks to the current material of choice, steel. Some of the main benefits of using aluminum over steel are that aluminum is much lighter, more ductile, similar mechanical properties as steel, and, theoretically, 100% recyclable (melting down the scrap aluminum only requires 5% the energy used to produce aluminum from ore). Additionally, aluminum is easily machined, cast, drawn, and extruded, has excellent corrosion resistance properties, is a good electrical conductor as well as a good thermal conductor. Despite all these benefits, the limitations of aluminum are that it is more costly than steel, has lower fatigue strength and creep resistance (at elevated temperatures), has high shrinkage and, consequently, high shrinkage porosity, and, lastly, is susceptible to hot cracking that causes failure.

2.3 Aluminum Trends

Aluminum has been used in the aerospace industry ever since 1903 when the Wright brothers used it in their historic airplane. Eighty percent of modern commercial aircrafts consist of aluminum by weight. Aluminum has also been used for marine applications and rail cars to increase speed as well as fuel economy. In addition, aluminum has been used as a packaging material due to its easy manufacturing and corrosion resistance. Aluminum is also readily used in commercial and residential constructions as window and door frames, flashing, gutters, downspouts, as well as roofing. Despite numerous applications in all of these industries, the automotive industry has not used aluminum to the same extent. Many automotive companies (especially in Europe) have turned to aluminum to innovatively reduce fuel consumption and CO_2 emissions. The European automotive industry has more than doubled the average amount of aluminum used in passenger cars during the last decade and plans to continue to increase the aluminum content.

Aluminum is a trending material for automotive usage but it does have distinct challenges, the two main challenges being joining and surface treatment. More recently, aluminum castings, extrusions, and sheet metal have been increasingly used for applications, such as engine blocks, powertrain parts, radiators, and some structural components. Audi has begun to use aluminum for the car frames in addition to body panels [4].

According to Ducker Worldwide studies, by 2025, more than 75% of pickup trucks and 20% of SUVs and large sedans produced in North America will be aluminum-bodied [5]. Additionally, according to Department of Energy (DOE) studies, reducing vehicle weight with aluminum can result in the lowest total vehicle lifecycle environmental impact as compared to traditional and advanced steels.

In 2015, the total aluminum content for the 17.46 million expected production vehicles will increase to nearly 7 billion pounds [3]. Of these 7 billion pounds, 11% is expected to be used in body and closure components, and 33% is expected to be used for engine parts.

References

1. Materials Selection in Mechanical Design by Ashby Elsevier, UK., 5th edition, 2018; https://grantadesign.com/education/students/charts/
2. Material Selection for Mechanical Design-I, II and III by Gregory Kirchain Kirchain, MIT, 2005. Lecture slides are available at: https://www.yumpu.com/en/document/read/8771102/materials-selection-for-mechanical-design-i

   https://ocw.mit.edu/courses/materials-science-and-engineering/3-080-economic-environmental-issues-in-materials-selection-fall-2005/lecture-notes/lec_ms3.pdf
3. The Automotive Body: Volume 1: Components Design by L. Morello, et al., Springer, 2011.
4. Young's Modulus – Density *available at:* www-materials.eng.cam.ac.uk/mpsite/interactive_charts/stiffness-density/NS6Chart.html
5. "Automotive Trends in Aluminum, The European Perspective: Part Two." Stress Corrosion Cracking of Aluminum Alloys, 2005, available at: www.totalmateria.com/page.aspx?ID=CheckArticle&site=ktn&NM=137

3

Mechanical and Physical Properties of Aluminum

3.1 Stress and Strain Considerations

Stress and strain are the fundamental considerations for any mechanical application. The stress-strain curve for materials is developed by loading test specimens (often called dog bone specimen) in accordance with ASTM or DIN procedures. The stress-strain curve can be used to read off the structural load-bearing capability (sometimes called loadability) of a certain material. Loadability is one of the most critical material properties when it comes to selecting the most suitable material for an application. Loadability is tested in a standardized tensile test with a procedure defined by ASTM or DIN standards (Figure 3.1). The overall length of the plate specimen is usually around 8 inches and width 0.5 inch. The thickness is about 0.005 inch. The gage length is 2 inches, and the width in the gap section is about 0.75 inch. The basics of the test include slowly and steadily pulling the test specimen in tension until the specimen fractures under the applied load. Different materials behave in different ways. So, using this standardized test, engineers can compare different materials based on each material's stress-strain curve (test results).

Figure 3.2 is an example of a stress-strain curve for a brittle and ductile material. The figure also shows the two main regions for the ductile material: the elastic region and the plastic region. Engineers design components with materials that remain in the elastic region. Once a component reaches the plastic region, it can no longer perform its intended function safely. When a material is loaded and the total stress does not exceed the yield stress, the material will return to the original form according to Hooke's Law. This phenomenon is called elastic behavior. In the elastic region, the load versus deformation (or the strain-strain) curve will be linearly approximated according to Hooke's Law (Equation 3.1). Thus, Hooke's law states that tensile stress is directly proportional to the strain by the modulus of elasticity (E). The modulus of elasticity is the slope of the stress-strain curve in the elastic region. For materials, such as aluminum and steel, the modulus of elasticity

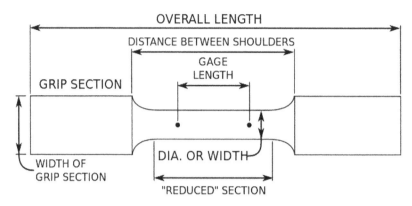

FIGURE 3.1
Tensile test specimen [1].

can be assumed to be constant. The same is not true for materials, such as polymers and some composites.

$$\sigma = E * \varepsilon \qquad (3.1)$$

If the applied loading causes the material to experience stress above the yield point, the material will not return to its original form and will

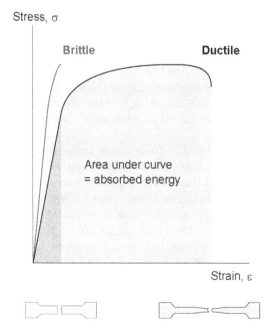

FIGURE 3.2
Stress-strain curve for ductile material [1].

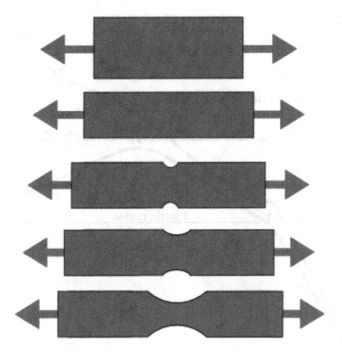

FIGURE 3.3
Necking of test specimen [2].

experience some form of deformation. The region after the yield stress is called the plastic region. Yielding is the point where deformation starts to occur. This region is where the strain increases with a very little increase in stress. Another region is the strain-hardening region where the strain continues to increase with a small increase in stress. During this time, the material gets harder due to the movement of dislocations (irregularity or defect within a material's crystallographic structure). As the loading continues to increase, necking begins to occur as shown in Figure 3.3 for a flat specimen. Necking is where a component suddenly deforms and fracture occurs.

There are two types of stress-strain curves: engineering stress-engineering strain and true stress-true strain curves (Figure 3.4). Engineering stress-engineering strain curves are a somewhat idealized version of the stress-strain curve. The engineering stress (or nominal stress) uses the original cross-sectional area of the specimen and the original force applied to the specimen (Equation 3.2). Additionally, engineering strain is calculated using the final and original lengths (Equation 3.3). On the other hand, true stress is calculated using the instantaneous force (not original force) and instantaneous cross-sectional area (not original area) (Equation 3.4). Moreover, the

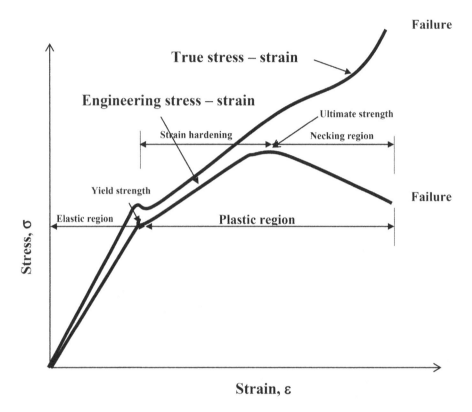

FIGURE 3.4
Engineering stress-engineering strain vs. true stress-true strain curves (adapted from ref [3]).

strain is calculated using the instantaneous change in length divided by the instantaneous length (Equation 3.5).

$$\sigma_n = \frac{F}{A_o} \tag{3.2}$$

Where σ_n is the nominal stress, F is the nominal force, and A_o is the original cross-sectional area.

$$\varepsilon_n = \frac{L - L_0}{L_0} \tag{3.3}$$

Where ε_n is the nominal strain, L_0 is the original length, and L is the final length.

$$\sigma_{true} = \frac{F_i}{A_i} \tag{3.4}$$

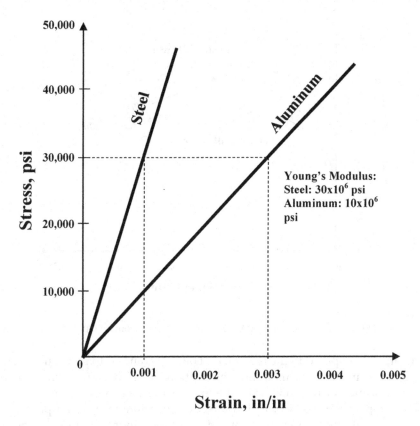

FIGURE 3.5
Comparison of elastic behavior of steel and aluminum (not to scale) – (adapted from ref [3]).

Where σ_{true} is the true stress, F_i is the instantaneous force, and A_i is the instantaneous cross-sectional area.

$$\varepsilon_{true} = \frac{L - L_i}{L_i} \tag{3.5}$$

Where ε_{true} is the true strain, L_i is the instantaneous length, and L is the final length.

Steel is the most commonly used material for many automotive applications. However, aluminum and its alloys are beginning to be used as a replacement for steels for certain applications. Figure 3.5 illustrates the elastic behavior of steel versus aluminum. It can be seen that for a given stress, aluminum deforms elastically three times as much as steel. Additionally, when compared to steel sheets, aluminum sheets (5xxx and 6xxx series) exhibit tensile stress and tensile strength nearly identical to those of mild steel (Figure 3.6). Despite this, aluminum with these properties has smaller

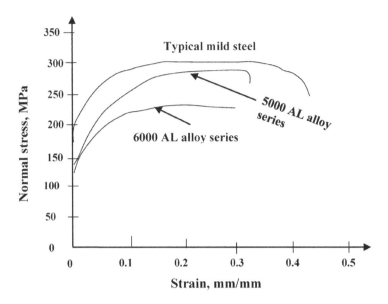

FIGURE 3.6
Stress-strain diagrams for (a) steel; and (b) aluminum (adapted from ref [3]).

elongations compared to the mild steel. This fact comes into play when forming aluminum sheets. This limited elongation causes defects, such as wrinkles, tears, and spring-back, when manufacturing sheet aluminum.

The temperature has a profound effect on aluminum's material properties
(Figure 3.7 (a)). Increasing the temperature of aluminum reduces the tensile

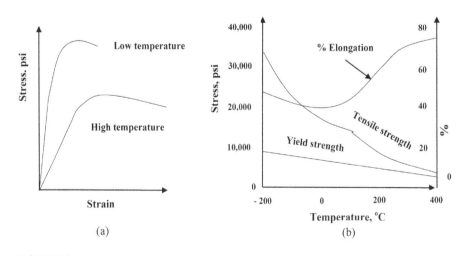

FIGURE 3.7
The effect of temperature on (a) aluminum stress-strain curve; and (b) tensile properties of
aluminum alloy (not to scale) – (adapted from ref [4]).

strength and yields strength while increasing the percent elongation and strain. Higher temperatures make aluminum a more ductile material where low temperatures make aluminum behave like a more brittle material. Taking the operating temperature of a component into account is crucial for the success of the part. If the part is operating in hot conditions, the designer needs to take into account the change in properties in order to make a safe component (Figure 3.7 (b)).

3.2 Rigidity and Stiffness

Modulus of rigidity (G) is a measure of how easily a material can be twisted or sheared. The modulus of rigidity is similar to the modulus of elasticity in the sense that the modulus of elasticity is the slope of the elastic region of the tensile stress-strain curve where the modulus of rigidity is the slope of the shear stress-strain curve. Engineers consider this property when designing parts or protocols such as torque specifications of driveshafts, or anything that will experience a torsional load during operation.

Stiffness is a measure of how much force or moment loading is needed to stretch (elongation or compression), bend, or twist through and angle of a test specimen. Designers consider this material property when designing members that will experience some combination of linear and angular deflections during normal operation. A classic example of where stiffness is incredibly important is when designing an automobile chassis frame. The value of stiffness is a function of loading (axial, torsional, or bending). Axial Stiffness is determined by measuring the load and the consequent deformation as shown in the schematic of Figure 3.8. The axial stiffness test will generate an axial stiffness constant (k) which is calculated by dividing the total load by the total elongation and has units of force per distance (N/m, lbf/in, etc.). Bending stiffness is determined by measuring the load and the bending deflection it causes. Determining the bending stiffness theoretically is more complicated than it might seem. The bending

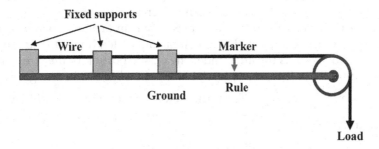

FIGURE 3.8
Schematic of an axial stiffness test apparatus.

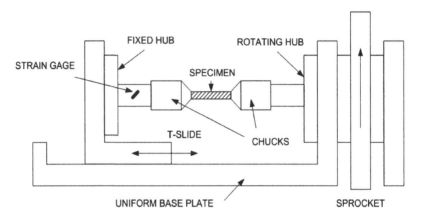

FIGURE 3.9
Torsion stiffness apparatus.

stiffness depends on how, where, and types of loads on the beam (simply supported beam, cantilever beam, distributed load, point load, etc.). In order to get the bending stiffness, it is often experimentally calculated. However, doing this for real-life applications can be challenging as well. Advanced analytical tools, such as computer-aided engineering (CAE) and finite element analysis (FEA), are needed and this entire testing can be very expensive. Determining torsional stiffness is easier to do with experiments than a theoretical calculation. The theoretical calculation is similar to the calculation for bending stiffness where it requires FEA. The experimental method involves a torsion stiffness apparatus that applies a known torque to the sprocket and twists the test material until failure as shown in the schematic of Figure 3.9.

3.3 Resilience, Toughness, and Strength-to-Weight Ratio

The modulus of resilience is a mechanical property of a material that shows how effective the material is in absorbing mechanical energy without exceeding the elastic limit. This mechanical property is important when designing components that are subjected to shock or impact loading. The modulus of resilience is defined as the area under the elastic portion of the stress-strain curve (Figure 3.10). It is a property that is related to strain energy density with units of N/m^2 or J/m^3. The formula for the modulus of rigidity is given in Equation 3.6 [3, 4]

$$U_r = \frac{\sigma_y^2}{2 * E} \tag{3.6}$$

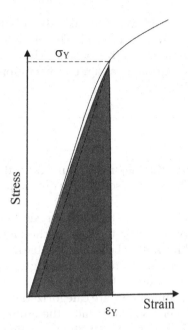

FIGURE 3.10
Definition of modulus of resilience.

In the above equation, U_r is the modulus of rigidity, E is the modulus of elasticity, and σ_y is the yield stress.

Modulus of toughness is a mechanical property that indicates a material's ability to handle overloading conditions prior to the fracture limit. This mechanical property is an indicator of how much energy a material can absorb before the material itself fractures and fails. A simplified formula of the modulus of toughness is given by Equation [3.7]

$$T_m = 0.5 * \left(\sigma_y + \sigma_u \right) * \varepsilon_{fracture} \qquad (3.7)$$

In the above equation, T_m is the modulus of toughness, σ_y is the yield stress, σ_u is the ultimate stress (maximum stress), and $\varepsilon_{fracture}$ is the fracture strain.

Strength-to-weight ratio is given by Equation [3.8] used to describe the measure of efficiency in a structure and is normally associated with large things such as chassis frames. The strength-to-weight ratio compares the weight of the structure to the amount of weight it can bear without failing. Simply put, the strength-to-weight ratio is the strength of a structure divided by the total weight of the structure.

$$Strength - to - Weight\ Ratio = \frac{Structure\ Strength}{Structure\ Weight} \qquad (3.8)$$

Compared to steel, aluminum is typically about half as strong but it also measures a third of the weight of steel. Therefore, the strength-to-weight ratio of aluminum is higher than that of steel. This is one of the big reasons why aluminum is being turned to as a replacement for steel in automobiles.

3.4 Aluminum Alloys

This section will discuss the major metallurgical differences between various aluminum alloys. Table 3.1 describes the typical properties of aluminum [3, 4].

Pure aluminum is a soft, ductile, and corrosion-resistant material with high electrical conductivity. Given these properties, it is no surprise why aluminum is widely used for foil and conductor cables. However, alloying with other elements is paramount when higher strengths or other properties are necessary for a given application. Different alloying elements give different properties. These aluminum alloys are categorized into series depending on the major alloying element which dictates the aluminum alloy properties. Table 3.2 briefly introduces what major alloying elements correspond to the different aluminum alloy series.

TABLE 3.1

Typical Properties of Aluminum

Property	Unit(s)	Value
Atomic number	X	13
Atomic weight	g/mol	26.98
Valency	X	3
Crystal structure	X	FCC
Melting point	°C	660.2
Boiling point	°C	2480
Mean specific heat (0–100 °C)	$cal/g * °C$	0.219
Thermal conductivity (0–100 °C)	$cal/cms * °C$	0.57
Coefficient of linear expansion (0–100 °C)	$1/°C$	0.0000235
Electrical resistivity at 20 °C	$\mu\Omega * cm$	2.69
Density	g/cm^3	2.6898
Modulus of elasticity	GPa	68.3
Poisson's ratio	X	0.34

TABLE 3.2

Designations for Alloyed Wrought and Cast Aluminum Alloys

Major Alloying Element	Wrought Series	Cast Series
None (99% + aluminum)	1XXX	1XXX0
Copper	2XXX	2XXX0
Manganese	3XXX	
Silicon	4XXX	4XXX0
Magnesium	5XXX	5XXX0
Magnesium + silicon	6XXX	6XXX0
Zinc	7XXX	7XXX0
Lithium	8XXX	

Different alloying elements give the aluminum alloy certain properties. These different properties could be increased strength, increased machinability, increased weldability, etc. However, these alloying materials can make the aluminum alloy less conductive, less resistant to corrosion, etc. Table 3.3 represents a list of various alloys and their applications where *S* stands for sheet, *P* stands for plate, and *E* stands for extrusion.

For most automotive applications, 5xxx and 6xxx series alloys are utilized. For all wrought aluminum alloys, there are two distinct categories: those which derive their mechanical properties from work hardening and those which depend on heat treatment and age hardening to determine their mechanical properties. 1xxx, 3xxx, and 5xxx series alloys typically have their properties determined by cold work usually done by cold rolling. The properties depend entirely on how much cold work is done and if any annealing or heat treatment is done after the cold work. There is specific nomenclature used to categorize each different material depending on how much cold work was done and if there was any thermal treatment. These are shown in Tables 3.4 and 3.5.

In Table 3.4, there are three types of strain-hardened codes: H1x, H2x, and H3x. H1x alloys are strain hardened only to obtain the required strength without any additional heat treatment. H2x alloys are strain hardened and partially annealed. This H2x designation also applies to products which are strain hardened more than the required amount and then reduced in strength (by annealing) to the desired strength. For alloys that age-soften, the H2x tempers have the same minimum ultimate tensile strength as similar H3x tempers. For other alloys, the H2x tempers have the same minimum ultimate tensile strength as similar H1x tempers but with somewhat elevated elongation. H3x alloys are strain hardened and then stabilized. The H3x designation applies to products which are strain hardened and then the mechanical properties are stabilized with low-level heat treatment. This stabilization

TABLE 3.3

Common Aluminum Alloys, Their Characteristics, and Uses

Alloy	Characteristics	Common Uses	Form
1050/1200	Good formability, weldability, and corrosion resistance	Food and chemical industry	S,P
2014A	Heat-treatable, high strength, nonweldable, poor corrosion resistance	Airframes	E,P
3103/3003	Nonheat-treatable, medium strength (work hardened), good weldability, formability, and corrosion resistance	Vehicle paneling, marine applications, mine cages	S,P,E
5251/5052	Nonheat-treatable, medium strength (work hardened), good weldability, formability, and corrosion resistance	Vehicle paneling, marine applications, mine cages	S,P
5454	Nonheat-treatable, good weldability and corrosion resistance, used at temperatures from 65–200 °C	Pressure vessels and road tankers, chemical plants	S,P
5083/5182	Nonheat-treatable, good weldability and corrosion resistance, resistant to seawater, and the superior alloy for cryogenic use (when annealed)	Pressure vessels and road transport applications below 65 °C and ship-building structures	S,P,E
6063	Heat-treatable, medium strength, good weldability, and corrosion resistance	Architectural extrusions, window frames, irrigation pipes, and intricate profiles	E
6061/6082	Heat-treatable, medium strength, good weldability, and corrosion resistance	Stressed structural members, bridges, cranes, roof trusses, and beer barrels	S,P,E
6005A	Heat-treatable, air-quenchable, not notch-sensitive, similar properties to those of 6082	Thin-walled, wide extrusions	E
7020	Heat-treatable, naturally age hardens (recovery of properties in welding zone over time), susceptible to stress corrosion, and good ballistic deterrent properties	Armored vehicles, military bridges, and motorcycle frames	P,E
7075	Heat-treatable, very high strength, nonweldable, and poor corrosion resistance	Airframes	E,P

typically improves ductility and is only applicable to alloys which gradually soften with time at room temperature.

In addition to nomenclature being used to describe the degree of strain hardening, there is nomenclature to describe how aluminum alloys are

TABLE 3.4

Nomenclature for Work-Hardened Aluminum Alloys

Symbol	Description
O	Annealed, soft
F	As fabricated
H12	Strain hardened, quarter hard
H14	Strain hardened, half hard
H16	Strain hardened, three quarters hard
H18	Strain hardened, fully hard
H22	Strain hardened, partially annealed, and quarter hard
H24	Strain hardened, partially annealed, and half hard
H26	Strain hardened, partially annealed, and three quarters hard
H28	Strain hardened, partially annealed, and fully hard
H32	Strain hardened, quarter hard, and stabilized
H34	Strain hardened, half hard, and stabilized
H36	Strain hardened, three quarters hard, and stabilized
H38	Strain hardened, fully hard, and stabilized

TABLE 3.5

Heat Treatment Designations [Adapted from Reference 4]

Code	Description
T1	Cooled from an elevated temperature-shaping process and naturally aged to a subsequently stable condition
T2	Same as T1. Additionally, it is cold worked
T3	Solution heat treated, cold worked, and naturally aged to a subsequently stable condition
T4	Same as T3 but without the cold work
T5	Same as T1 but artificially aged
T6	Solution heat treated the artificially aged
T7	Solution heat treated the overaged/stabilized
T8	Same as T6 but cold worked before aging
T9	Same as T8 followed by cold work
T10	Cooled from an elevated temperature-shaping process, cold worked, then artificially aged

solution heat treated or age hardened. This is done by adding the letter *T* followed by a number to describe the various conditions.

When it comes to aluminum alloys, a vast amount of knowledge and understanding is required in order to process aluminum and how to use the different alloys for different applications [5]. Being familiar with all of this information can help engineers make informed decisions on material selection for any application, especially automotive applications.

References

1. Tensile specimen figure available at: https://commons.wikimedia.org/wiki/ File:Tensile_specimen.png, 2018.
2. Different stages of tensile testing of a specimen, available at: https://upload. wikimedia.org/wikipedia/commons/5/5c/Schematic_Diagram_of_Necking. png
3. The Science and Engineering of Materials, 4th Ed., by Donald R. Askeland and Pradeep P. Phule, Chapter 6, Brooks/Cole Publisher, 2003.
4. Hibbler R. C., Mechanics of Materials, 8th edition. Prentice-Hall 2010.
5. Alcoa Structural Handbook: A Design Manual for Aluminum, 1956.

4

Aluminum-Manufacturing Methods

4.1 Introduction

In this chapter, many different manufacturing processes are discussed briefly. The following sections are focused on their relation to aluminum due to aluminum being the primary material for lightweighting. Many parts and subsystems in a typical automobile are manufactured using conventional and advanced manufacturing processes and technologies. Automotive companies are turning their focus from steel to aluminum mostly due to aluminum's light weight and consequent energy savings. Due to aluminum gaining attention, it is necessary to understand the different manufacturing processes in order to identify which process is the most economical.

4.2 Machining

Machining is the process of taking raw material and making into some component. The most common machining operations are turning, drilling, reaming, milling, grinding, and threading (using a tap of the thread-rolling process). The most common machines used to perform the above processes are lathes, mills, and grinders. Figures 4.1 and 4.2 show a conventional lathe and a vertical milling machine, respectively.

There are many advantages of using aluminum as the stock material for automotive components. Aluminum is a much softer material than steel, making it easier to machine. The softness quality of aluminum and its alloys makes traditional machining operations (turning, milling, boring, tapping, sawing, etc.) easy to perform. A third benefit of aluminum being softer than steel is that the cutting force required to machine aluminum is substantially less. For the same section of chips, the force to machine aluminum is approximately one-third of that required for low carbon steel. It follows that for the same cutting force, chip removal is three times higher with aluminum alloys, such as 2017A whose mechanical properties are at par with low-carbon steel.

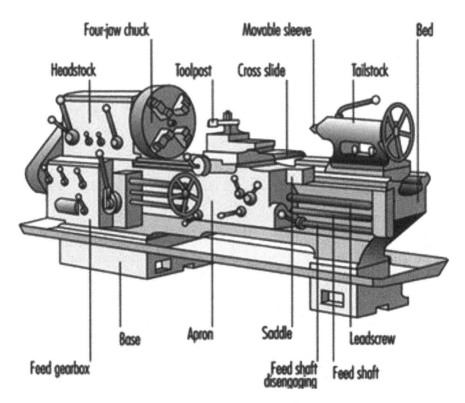

FIGURE 4.1
Lathe diagram [1].

The machines that are used to machine aluminum can be the same as for steel. However, for optimum aluminum machining conditions, rotational speed and feed rates can only be achieved on machines specifically designed for machining aluminum. Using aluminum in the machining process also allows for higher cutting speeds, better-machined surface finishes, and excellent dimensional control. In addition, longer tool life is experienced when machining aluminum over steel when best practices are followed. Additionally, aluminum's thermal conductivity assists heat transfer from the machining process to the chips. Unlike steel, the chips take the majority of the heat away from the workpiece and there is no need to provide heat treatment of the "stressfree annealing"-type during machining.

Despite all the advantages to machining aluminum over steel, there are drawbacks. Since aluminum's modulus of elasticity is around one-third that of steel, multiple jaw chucks or creative clamping arrangements are required to ensure that distortion or deformation of the aluminum stock does not occur. In addition, high-speed machining can result in the aluminum adhering to the cutting tool. When the aluminum chip adheres to the cutting tool, more heat is generated causing the tool life to be shortened, and the

FIGURE 4.2
View of Fresadora vertical milling machine [2].

surface finish is crappy because instead of the tool doing the cutting, the chip adhered to the tool is doing the cutting (the built-up edge also dulls the tool). Tool coatings and tool material selection are the two primary techniques to reduce the built-up edge in order to increase tool life. High-speed machining also degrades the cutting ability of the tool.

Aluminum has great potential for alloying. Due to the wide variety of aluminum alloys available, there are different machinability ratings for different alloys. Machinability ratings of aluminum are based on chipping characteristics. Table 4.1 describes the different ratings.

In addition to different machinability ratings, different aluminum alloys have different characteristics which directly affect machining. Table 4.2 describes machinability characteristics of various aluminum alloys used in typical automotive applications.

TABLE 4.1

Aluminum Machinability Ratings [Adapted from Ref 3 and 4]

Rating	Characteristics
A-rated	Very small chips and excellent surface finish (2011 and 6020 alloys)
B-rated	Curled or easily broken chips and good to excellent surface finish (2024, 2017, 4032, 6013, 6262, and 7075 alloys)
C-rated	Continuous chips and good surface finish (6005 and 6061 alloys)
D-rated	Continuous chips and satisfactory surface finish (5056 and 6063 alloys)

A turning operation is an operation where the workpiece or the tool is rotated to remove material to manufacture a finished component (usually performed with a lathe or mill). Mills and lathes are very suitable for aluminum alloys. Milling uses a rotating tool that continuously brings fresh edges into action. Lathes use a stationary tool that cuts a rotating workpiece where the same cutting edge is continuously in use. For both mills and lathes, the feeds and speeds are incredibly important.

TABLE 4.2

Machinability Characteristics of Various Aluminum Alloys [Adapted from Various References, Example, Ref 3 and 4]

Alloy	Machinability Characteristics
2011 alloy (tempered: T3, T451, T8)	Free machining (Pb and Bi addition)
	Readily available
	Poor anodizing response
	Poor corrosion resistance
	Cold finished only
6262 alloy (tempered: T6, T651, T8, T9)	Intermediate machinability (Pb and Bi addition)
	Better corrosion resistance and anodizing response than 2011/2024 alloys
	Available in cold finished and extruded product
2024 alloy (tempered: T351, T4, T6)	Excellent deep drilling characteristics
	Improved stress corrosion cracking (SCC)
	Available in cold finished and extruded product
6061/6082 alloy (tempered: T6, T6511)	Mostly used alloy by automotive designers
	Moderate machinability
	Chipping is not as good as 6262 alloy
	Good corrosion and anodizing response
7075 alloy (tempered: T6, T6511, T73, T7351)	High-strength alloy for bumper and aerospace applications

To summarize in brief, there are a number of different feeds and speeds that depend on the tooling as well as the alloy being turned. Speed (when talking about a machining process) refers to the rotational speed of the workpiece (on a lathe) or the tool (in a mill). Feed refers to the linear speed that the tool moves to cut away material. The combination of the correct feed and correct speed will yield a great surface finish. If the feed or speed is too fast or too slow, the surface can come out with a poor finish or worse can affect the longevity of the tooling; see, for example, reference 4 for details. Tooling also affects feeds and speeds. There are different tooling characteristics depending on what alloy is being machined. As an example, for 2011, 6020, and 6262 alloys, the top rake angle should be less positive for free cutting alloys (0–5 degrees). For 2024, 6061, 6082, and 7075 alloys, the top rake angle should be more positive for B and C-rated materials (5–10 degrees). These are shown in Figure 4.3.

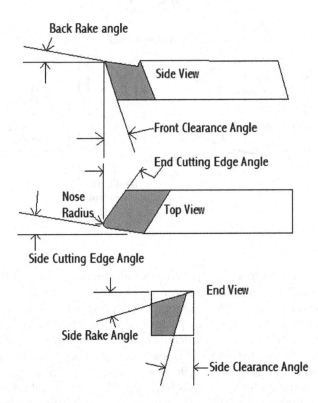

Tool Bit Geometry

FIGURE 4.3
Typical hand ground-cutting tool angles for the lathe [5].

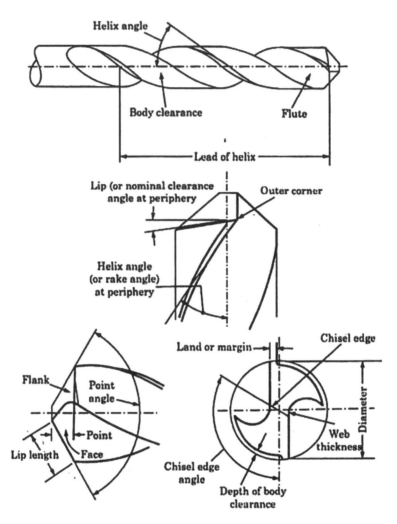

FIGURE 4.4
Cutting tool geometry of a typical twist drill [6].

Another important machining process for aluminum is drilling (Figures 4.4 and 4.5). Drills for machining aluminum typically have large helix angles and large, polished flutes for rapid chip removal to prevent build-up. Unlike turning and milling, drilling is usually performed with cutting fluids, like mineral oils, emulsions, or aqueous chemical solutions. Standard drills (for drilling steels) can be used to drill shallow holes; however, specialty drills specifically for aluminum are available. Despite aluminum being considered one of the easiest machining materials, it does have some issues when it comes to drilling. The most commonly drilled aluminum alloys in the automotive industry are 6061 and 7075. Due to aluminum being very ductile and soft, the chip from

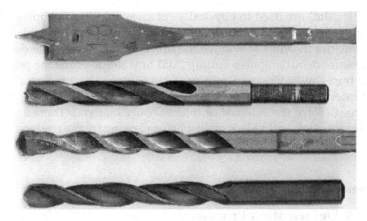

FIGURE 4.5
Examples of spade, lip and spur (brad point), masonry bit and twist drill bits [7].

drilling is in contact with the rake face for longer periods of time causing the bits to dull. In addition to this, holes drilled in aluminum may become over-sized. In soft alloys, the diameter may be slightly larger than the drill. If tight tolerances are required, smaller drills are typically used. The holes are reamed out later on for more precision.

Similarly, to other machining operations, there are best practices for drill-ing. For general purpose-drilling, 25–30-degree helix angle is used, whereas for deep-hole drilling, a 40–50-degree helix angle is used. To minimize chip packing and build-up, highly polished, parabolic, symmetrical flutes are used. Lastly, for deep-hole drilling, 130–140-degree point angle is used to clear chips easily.

Threading, tapping, and thread rolling are all machining processes where threads are made in order for the part to thread into another part or for another part like a bolt, to be threaded into the part. Threading is where special dies that have the proper rake and chamfer angle are used for cut-ting external threads onto a part. Tapping is where a special tap is used for cutting an internal thread into a part. Taps for aluminum should have deep, well-polished flutes. Spiral fluted taps are used for blind holes where straight fluted taps are used for through holes. Lastly, thread rolling is where the material is rolled on a special tool to press threads into a material forming external threads. Thread rolling is faster than using a die (for threading) and the rolled threads are stronger and have improved fatigue resistance than cut threads.

For nearly every machining process, the cutters produce chips that have not completely been cut away from the workpiece; these are called burrs. Before the workpiece is finished and can be put into service, these burrs must be removed. This process is called deburring. There are many ways to deburr workpieces. Mechanical deburring is where manual tools (sanding blocks, files, etc.) are used to physically cut away the external burrs. Tumble deburring involves putting the workpiece into a drum and is tumbled among abrasive media in order to remove the burrs. Similar to tumble deburring, high-pressure deburring uses cutting fluid or water to erode the protruding external burrs away. Thermal deburring uses intense heat to vaporize internal and external burrs. Lastly, chemical deburring is where a strong chemical is used to dissolve the external or internal burr material. There are several online information and videos available for viewing by the readers.

4.3 Bulk Deformation Processes

Bulk deformation is metal casting or forming operations which cause significant shape change by deformation in metal parts whose initial form is liquid (casting) or bulk, rather than sheet. Typical starting forms for bulk deformation processes are cylindrical bars and billets and rectangular billets and slabs. There are numerous bulk deformation processes that are going to be discussed in this section. These processes work by stressing the metal sufficiently to cause plastic flow into the desired shape. There are advantages and disadvantages to each process and will be discussed in depth in the following subsections. Bulk deformation processes are important because in hot working, significant shape change can be accomplished and there is little to no waste. When using a bulk deformation process, some operations make near net parts that require little or no machining. Cold worked parts increase their strength during the deformation process.

4.3.1 Casting

The automotive industry is the largest market for aluminum casting. More than half of the aluminum used in cars are components that were made using casting. Cast aluminum transmission housings and pistons have commonly been used in automobiles since the early 1900s. Most auto parts made of aluminum are cast with accurate and systematic working. These are subsequently machined with the processes discussed above. Common automotive parts that are fabricated using casting include:

- Aluminum wheels
- Aluminum transmission housings
- Engine blocks
- Wheel spacer components
- Carburetor housings

- Aluminum brackets
- Alternator housings

Casting is a simple, inexpensive, and versatile way of forming aluminum into a wide array of products. Casting is the original and most widely used method of forming aluminum into products. Technical advances have been made but the fundamental process has remained the same; molten aluminum is poured into a mold to duplicate the desired pattern. Casting molds must be designed to accommodate each stage of the process. For part removal, a slight taper called the draft must be used on surfaces perpendicular to the parting line so the pattern can be removed from the mold when the molten aluminum has solidified. Cavities in the final product can be tricky problems to solve. To produce cavities within castings, negative forms are used to make cores. These cores are what make the cavity. Casts of this nature are usually produced in sand molds where the cores are inserted into the casting box after the pattern is removed. Aluminum's properties of light weight and strength bring fundamental advantages when cast into parts. The three most important methods of casting are:

1. Die casting
2. Permanent mold casting
3. Sand casting

The die casting process forces molten aluminum into a steel die (mold) under pressure. Die casting is normally used for high-volume production and forms precise aluminum parts that require minimal machining and finishing. When using die casting to form aluminum, the final weight is an important consideration. Die casting has the maximum weight limit of approximately 70 pounds for aluminum. The thickness of the final product is important as well. The minimum part thickness the die casting aluminum can support is roughly 0.035 inch and the maximum part thickness should be less than 0.5 inch. One common application of die-cast aluminum is thin-celled enclosures with ribs and bosses on the interior to maximize strength.

There are two main processes of die casting: hot-chamber (or gooseneck, Figure 4.6) and cold-chamber die casting (Figure 4.7) [ref 8]. For both methods, a piston is used to deliver liquid metal into the die. The main difference between the two methods is that for hot-chamber die casting, the piston is inundated in a vat of molten aluminum, whereas in the cold-chamber method, the liquid metal is delivered one "shot" at a time to the piston. The cast is held under pressure during the solidification process for both methods.

The advantage of die casting as opposed to other methods of casting is that intricate detail and the best surface finish can be achieved of any of the casting methods. This surface finish and detail is due to the high pressures employed during the solidification process coupled with the smooth die

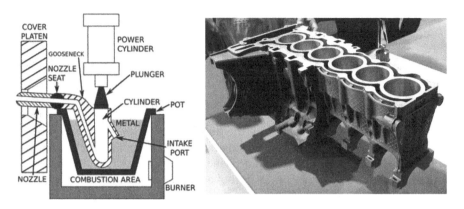

FIGURE 4.6
Schematic of hot-chamber die casting machine (left); an engine block with aluminum and magnesium die castings (right) [8].

surface. Another advantage is that die casting promotes extremely fast production. Unlike sand or permanent mold casting, die casting allows for thin casting walls as well as have cast-in inserts, such as threaded inserts, heating elements, and high strength-bearing surfaces. Despite the advantages, there are limitations to die casting. The biggest limitation is that the steel dies are not permeable. Vents are required for gases to escape and care must be taken to ensure that the pressurized molten aluminum does not escape through

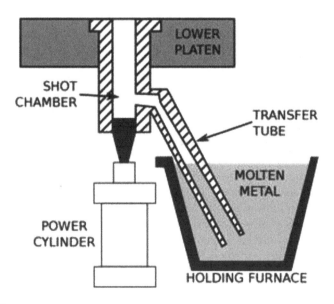

FIGURE 4.7
Schematic of cold-chamber die casting [8].

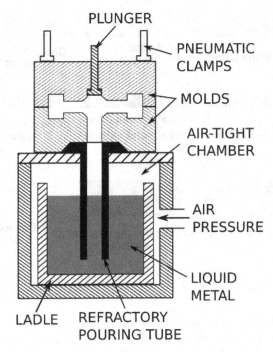

PLUNGER

PNEUMATIC CLAMPS

MOLDS

AIR-TIGHT CHAMBER

AIR PRESSURE

LIQUID METAL

LADLE   REFRACTORY POURING TUBE

FIGURE 4.8
Schematic of the low-pressure permanent mold casting process [9].

the vents. High melting-point metals cannot be die-cast because the molds cannot withstand the high temperatures. Additionally, the initial cost of getting a die-cast-manufacturing operation set-up is high due to the high cost of dies and any automation features. Another disadvantage of die casting is that the final casting will have a very small amount of porosity that prevents any heat treating or welding which causes microcracks inside the part and exfoliation of the surface. Lastly, thermal fatigue is a concern for the molds used in die casting.

The second casting process is permanent mold casting, also commonly known as gravity die casting (Figure 4.8). Permanent mold casting involves molds and cores of steel, bronze, refractory metals, and graphite. Molten aluminum is usually poured into the mold and, on occasion, a vacuum is applied. To improve the life of the mold, the surface of the mold can be coated with a refractory slurry and the mold is heated to reduce thermal fatigue. Permanent mold castings can be made stronger than sand or die castings. Some common parts that are manufactured using permanent mold casting include gears, gear housings, pipe fittings, pistons, impellers, and wheels [10].

The advantage of permanent mold casting is that the cooling is faster than that of a sand casting which leads to finer grain structure, reduced porosity and higher overall strength. The mold for permanent mold casting is reusable

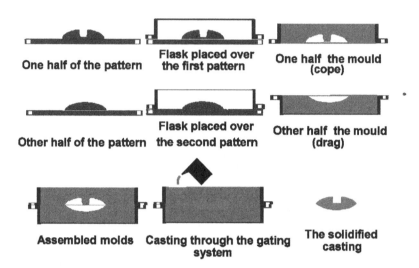

One half of the pattern   **Flask placed over the first pattern**   **One half the mould (cope)**

Other half of the pattern   **Flask placed over the second pattern**   **Other half the mould (drag)**

Assembled molds   **Casting through the gating system**   **The solidified casting**

FIGURE 4.9
Schematic of sand casting process [11].

and yields parts with good finishes and high production rates. Additionally, retractable metal cores can be used to create undercuts while maintaining a quick action mold. Lastly, directional solidification can be induced by changing the mold wall thickness or by directly heating or cooling certain portions of the mold. The main disadvantages of permanent mold casting are the high costs of molds, only low-melting-point metals can be used, and the molds have a fairly short life due to thermal fatigue and erosion. Other disadvantages of permanent mold casting are that similar to die casting, the molds are not permeable and vents are required for the gases to escape. Lastly, the mold is rigid. Therefore, hot tears can be created during solidification caused by shrinkage if the cast is not quickly removed.

Sand casting is the most versatile method for producing aluminum products (Figure 4.9). The process starts with a pattern that is a replica of the finished casting. Virtually, any pattern can be pressed into a fine sand mixture to form the mold into which the molten aluminum is poured. The pattern is always larger than the final part due to shrinkage and the need for finishing processes. Compared to die or permanent mold casting, sand casting is a slow process but more economical for small quantities, intricate designs or when very large castings are required. As shown in Figure 4.9, usually, there are six steps in this process [11]. These are:

1. A pattern is placed in sand to create a mold.
2. Both the pattern and sand are incorporated into a gating system.
3. The pattern is removed.

4. The mold cavity is filled with molten metal.

5. The metal is allowed to cool.

6. Lastly, the sand mold is broken away and the casting is removed.

Advantages of sand casting are that sand as a packing material is reusable, low-cost, permeable, temperature resistant, and nonreactive. Complex and intricate designs can be made with the use of cores. This also makes sand casting an attractive method when low volumes of parts are required due to the destruction of the mold after each part is made. Sand casting can be used to make a wide variety of parts from very small to very large or heavy. Lastly, metals with high melting points can be used. The limitation of using sand casting as a manufacturing process is that the surface is rough compared to die casting or permanent mold casting. Another disadvantage is that after each part is made, the mold is destroyed making this process unsuitable for mass production. Lastly, the mold is very susceptible to erosion when the molten metal is poured into the mold.

There are other forms of casting, such as shell molding, vacuum molding, plaster molding, and ceramic mold casting that are not covered in this book. However, another form of casting that is widely used in the automotive industry is called lost foam casting. The lost foam casting process originated in 1958 when H.F. Shroyer was granted a patent for a cavity-less casting method, using a polystyrene foam pattern embedded in traditional green sand. The polystyrene foam pattern left in the sand is decomposed by the poured molten metal. The metal replaces the foam pattern, exactly duplicating all of the features of the original pattern. This process requires that a pattern be produced for every casting poured due to the patterns being dissolved ("lost") in the process. Generally, all ferrous and nonferrous materials, such as aluminum, can be successfully cast using the lost foam pattern and gating system must be decomposed to produce a casting. Metal-pouring temperatures above 1000 °F are usually required. Lower-temperature metal can be poured; however, part size is limited. In addition, low-carbon ferrous castings will require special processing. The basic steps for the lost foam casting manufacturing process are presented below.

1. A foam pattern and gating system are made using a foam-molding press. The final pattern is approximately 97.5% air and 2.5% polystyrene.

2. The foam pattern and the gating system are glued together to form a cluster of patterns.

3. The cluster is coated with a permeable refractory coating (to avoid sand erosion) and dried under controlled conditions.

4. The dried, coated cluster is invested in a foundry flask with loose unbonded sand that is vibrated to provide tight compaction.

5. The molten metal is poured onto the top of the gating system which directs the metal throughout the cluster and replaces the foam gating and patterns.
6. The remaining operations, such as shakeout, cutoff, grinding, and heat treatment, are performed similarly to other casting processes.

Lost foam castings can be produced with most metals in a large variety of sizes from very small to very large. Typically, a linear tolerance of ±0.005 inch per inch is standard for the lost foam casting process. This tolerance will vary depending on part size, complexity, and geometry. Subsequent straightening or coining procedures will often enable even tighter tolerances to be held on critical dimensions. Due to the permeable refractory coating that is applied to the foam pattern, an excellent smooth finish results. If surface finish is a critical cosmetic requirement, then surfaces can be targeted to maintain an exceptionally smooth finish. The castings made by this process are generally more expensive than forged parts, or parts made by other casting processes. The value inherent in the lost foam process versus other processes is seen in tighter tolerances, weight reduction, and as-cast features which all result in less machining and secondary process cleanup time. Many castings that require secondary operations, such as milling, turning, grinding, and drilling, can be made using the lost foam casting process with only 0.020″–0.030″ of machine stock. Lost foam castings are used for many critical applications including, engine heads, turbochargers, marine motors, high-pressure pumps, and valves. X-ray and soundness testing on lost foam castings for shrinkage shows characteristics comparable to other casting processes. In order to make the foam patterns, dies are required to shape the polystyrene. These dies are highly specialized and expensive but are more favorable compared to permanent and die-cast tooling.

An advantage of lost foam casting is that there are no parting lines, cores, or risers enabling less secondary processes. The polystyrene used for the foam patterns is cheap. Additionally, intricate and complex shapes with excellent surface finish can be realized in the cast. As mentioned above, a limitation for this process is that the die for shaping the polystyrene is expensive and highly specialized. Lastly, the foam patterns can be easily damaged due to the low strength of polystyrene.

4.3.2 Forging

Forging is a deformation process that involves the workpiece being compressed between two dies. Some common components that are made using forging processes are engine crankshafts, connecting rods, gears, aircraft structural components as well as turbine parts. There are four major forging classifications: hot forging, cold forging, impact forging, and press forging.

Hot forging is the most common operation that is used because of the significant deformation, the need to reduce strength and increase the ductility of the metal being forged. Cold forging is not as common; however, the benefit of this method is that the strength is increased due to strain hardening. Impact forging is where a forge hammer applies an impact load. Forging hammers are either gravity drop or power drop. Gravity drop hammers are large, heavy rams that provide the impact just by using gravity. Power drop hammers accelerate the ram using pressurized air or steam. Press forging is where a forge press applies a gradual pressure to deform the workpiece. This pressure is done with mechanical presses, hydraulic presses, and screw presses. Mechanical presses convert the rotation of a drive motor into linear motion of a ram. Hydraulic presses use hydraulics to actuate the ram. Lastly, screw presses use a screw mechanism to drive the ram and provide the pressure.

In addition to the different forging methods, there are three main types of forging dies: open die forging (Figure 4.10), impression die forging [12], and flashless forging [12]. Open die forging is when the workpiece is compressed between two flat dies, allowing the metal to flow laterally without constraint. Figure 4.10 shows two dies of an ingot that will be processed into a wheel. "Edging" and "fullering" are the intermediate processes of concentrating material using concave- or convex-shaped open dies. These processes are so called because they are usually carried out on the ends of the workpiece.

Impression die forging is when the die surfaces contain cavities or impressions that are imparted into the workpiece, thus constraining the metal flow. Flashing is created during this method. Lastly, flashless forging is where the workpiece is completely constrained in the die and no excess flashing is produced. Further details of these processes can be found in the literature such as in Reference [10].

Since open die forging involves the compression of the workpiece between two flat dies, the effect of friction is extremely important. If no friction occurs between the cylindrical workpiece and the die surfaces, then homogenous

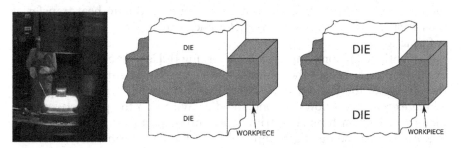

FIGURE 4.10
Example of an open die drop forging [12].

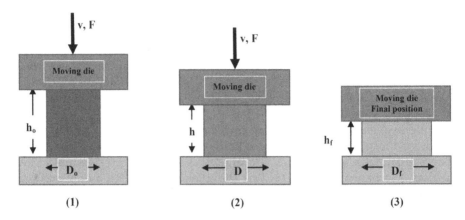

FIGURE 4.11
Ideal (frictionless) open die forging with homogenous deformation.

deformation occurs so that radial flow (in the horizontal position) is uniform throughout the workpiece height and true strain is given by Equation 4.1.

$$\varepsilon = \ln\left(\frac{h_0}{h}\right) \qquad (4.1)$$

Here, h_0 is the starting height and h is the height at some point during compression between the top-moving and the bottom-fixed dies. According to this equation, the true strain is at its maximum when $h = h_f$ where h_f is the final height. As seen in Figure 4.11 below, (1) is the start of the process with the workpiece at its original length and diameter, (2) is where the workpiece is under partial compression, and (3) is the final size of the part.

However, the more realistic open die forging scenario is with friction between the workpiece and die surfaces that constrain lateral (radial) flow of the workpiece resulting in a barreling effect. In hot open die forging, the effect due to friction is even more pronounced due to heat transfer at the die surfaces, which cools the metal and increases the resistance to the deformation.

As mentioned before, impression die forging [10] is where the compression of the workpiece causes the material to flow into cavities in the die that are the inverse of the desired part shape. Flash is formed during this process where excess metal flows beyond the die cavity into the gap between the die plates. This flash must be trimmed from the part later on but serves an important purpose during compression. As the flash forms, friction resists continued metal flow into the die gap, constraining the metal to fill the die cavity. Additionally, in hot forging, the metal flow is further restricted by cooling against die plates. When using impression die forging to make final components, several forming steps are usually required. The first steps are usually to redistribute the metal for more uniform deformation and the desired metallurgical structure in the following steps. The final steps bring the part to its final geometry. Like all processes, there are advantages and limitations. Some advantages would include high production rates, less material waste, greater strength, and favorable grain orientation. Some of the disadvantages would include the incapability of achieving close tolerances, and secondary operations like machining which are usually needed to form the final product. Similar to open die forging process, impression die forging is also carried in three or four steps from the initial position of the dies to the final position of dies before a green part is produced for further operations.

The last forging process is flashless forging [12], where the workpiece is compressed in a die with a punch where the die does not allow for a flash to form. For this process, the starting workpiece volume must equal the die cavity volume within a very close tolerance. This makes process control much more important than impression die forging. Flashless forging is best suited for simple geometries. As with the other forging processes, the flashless forging also involves multiple steps before a green part is produced.

4.3.3 Extrusion

Aluminum is the most commonly extruded material and can be extruded when hot (450–500 °C) or cold. There are two main types of extrusion processes: direct extrusion and indirect extrusion (Figures 4.12–4.17). Direct extrusion is the most common extrusion process and is also referred to as forward extrusion. Direct extrusion is the process where a cylindrical billet is placed in a chamber and forced through an extrusion die by a ram. The process is simple; however, the ram must overcome the frictional forces between the billet and the chamber. Indirect extrusion is where the die moves toward the billet and there is no relative motion between the billet and container. In the indirect extrusion process, there is negligible friction between

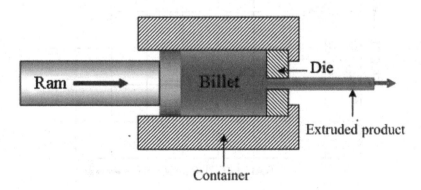

FIGURE 4.12
Schematic of direct extrusion method.

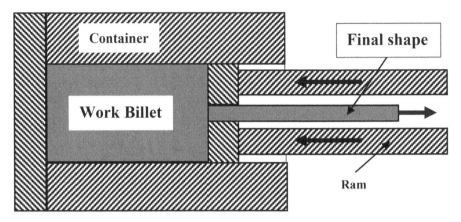

FIGURE 4.13
Schematic of indirect extrusion method.

the billet and the container because there is no relative motion between the two. This reduces the pressure of extrusion required in operations that typically have high billet-container friction. Despite this advantage, in indirect extrusion, the die is effectively the ram and is less rigid than a ram used for direct extrusion. This limits the amount of force that can be applied in the indirect extrusion process. Indirect extrusion is also more complicated than direct

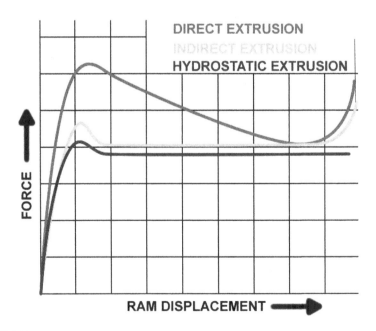

FIGURE 4.14
Plots of forces required by various extrusion processes [13].

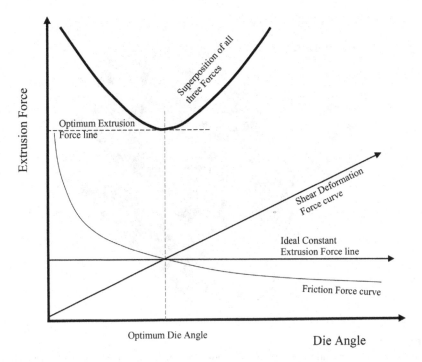

FIGURE 4.15
Plot of extrusion forces vs. die angle.

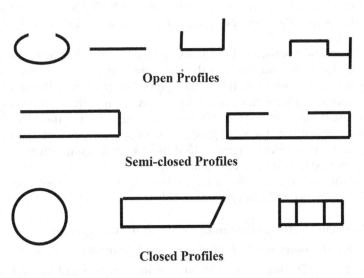

FIGURE 4.16
Basic extrusion shapes (Raghu adapted from Ref. [14]).

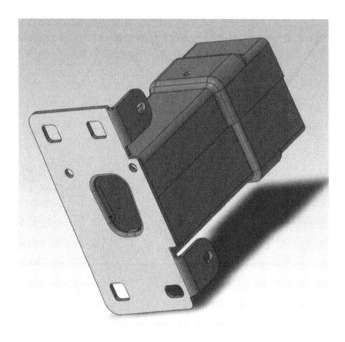

FIGURE 4.17
CAD image of an extruded aluminum crash box [reproduced with permission of Paris Velikis]
[15].

extrusion. Diagrams of both the extrusion processes as well as the extrusion forces with respect to ram displacement are shown in Figures 4.15 to 4.17.

Hot extrusion temperatures for aluminum can be 350 to 500 °C (650 to 900 °F), while for steel the corresponding temperatures are: 1200 to 1300 °C (2200 to 2400 °F).

The three main sources of extrusion resistance are the force due to friction, ideal force (no friction), and the shear deformation force. Friction force increases with the length of the contact patch between the billet and the die, and the contact path length increases with decreasing die angle. Shear deformation force increases with increasing die angle. Lastly, the ideal force is exclusively a result of the work to deform a homogenous material ignoring shear and friction forces. Figure 4.15 illustrates how qualitatively the three forces affect extrusion force based upon die angle.

Aluminum is used for extrusions due to its many advantages as compared to steel. When using aluminum over steel:

- The material thickness can be varied in the cross section
- Inner and outer webs and fins can be easily made
- Multichamber hollow profiles are made with standard technology
- Extrusion dies have longer lives and cost less

4.3.3.1 Types of Extruded Shapes, Aluminum Grades, and Applications

Aluminum extrusions come in different profiles and sizes suitable for automotive, aerospace, and other structural applications. As shown in Figure 4.16, there are four main types of profile shapes; open profile, semiclosed profile, closed profile, and solid profile. Solid profiles have low production and die costs and are not discussed due to their lack of automotive applications. For both open and semiclosed profiles, tooling is susceptible to a shorter life and has high material and die costs. Closed profiles can be very high cost (especially for multivoid hollow sections) in addition to high material and tooling costs.

Aluminum extrusions are extremely common in the automotive industry. The raw aluminum used for these extrusions is dependent upon the application. Some alloys are more suitable than others under certain conditions. The unalloyed 1xxx series is used for nonstructural applications and are the most suitable alloy for extrusion. 1xxx series alloys have high electrical and thermal conductivity, high corrosion resistance, excellent formability, finish capability, and extrudability, and are low-strength. 1xxx series applications might include tubes or equipment for electrical components, such as heat sinks and heat shields.

3xxx series alloys are low- to medium-strength and have good thermal conductivity, weldability, corrosion resistance, extrudability, and formability, but have poor machinability. Some applications of 3xxx series aluminum could include extruded tubes, radiators, heater cores, air conditioning evaporators, and coolers (automotive heat exchangers).

5xxx series alloys are medium- to high-strength with good corrosion resistance, weldability, machinability, and formability but have low extrudability.

6xxx series alloys are medium- to high-strength with good extrudability, corrosion resistance, weldability, machinability, and formability with the excellent surface capability and are used for some structural body parts. Some common applications for 6xxx series alloys are crash boxes, side-impact beams, seat components, bumper beams, engine cradle, space frame, and subframe. Figure 4.17 shows the CAD image of an example crash box. The most high-volume alloys in this class are 6060 and 6063 and are the most widely used extrusion alloy group. 6060 and 6063 are popular for high production in the automotive industry because their scrap values are decent, they can be delivered in temper T5, have lower Si and Mn compared to 6082, and can be heat-treated to acquire the desired grain structures and, consequently, the desired mechanical properties.

7xxx series alloys are high-strength alloys with low- to medium-extrudability, low- to fair-corrosion resistance, variable weldability, good machinability, and cold formability. 7xxx series alloys are typically used for bumper beams and the 7xxx alloys used for automotive applications differ from aerospace applications. 7xxx series have a fibrous grain structure which induces an increase in mechanical properties along the extrusion direction.

FIGURE 4.18
Extruded aluminum parts with several hollow cavities [16].

However, ductility may be reduced compared to a recrystalized grain structure. Examples of microstructure of 6xxx and 7xxx aluminum alloys can be found in reference [14]. The wall thickness of extruded parts can be constant or variable. An example of extruded aluminum with several hollow cavities and whose wall thickness is almost constant except at the ribbed areas at the center and corners is shown in Figure 4.18. The T-slots allow bars to be joined with special connectors.

4.3.3.2 Extrusions With Variable Wall Thicknesses, Webs, and Fins

In order to optimize cost efficiency, an extrusion profile design should be as production-friendly as possible. A symmetrical design with simple, soft-line, filleted corners with uniform wall thickness helps reduce cost. It is often acceptable to have a large range of wall thicknesses contained in a single profile. However, a profile with a uniform wall thickness is easier to extrude and, consequently, decreases die stress. See Figures 4.19 (a) and (b) for examples of open and closed section extrusions with variable wall thickness [17].

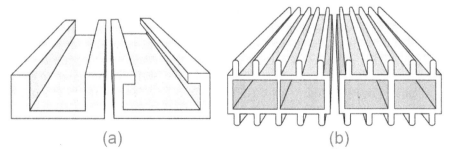

(a)   (b)

FIGURE 4.19
(a) Variable wall thickness vs. (b) Uniform wall thickness [17].

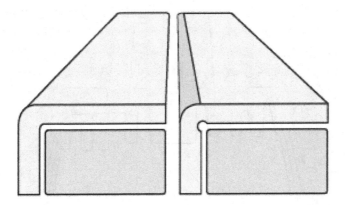

FIGURE 4.20
Sharp internal angles requiring unequal wall thicknesses [17].

However, there are a couple of important cases where nonuniform wall thicknesses occur. For strength reasons, it can be advantageous to concentrate the weight (thicker section) away from the center of gravity or to accommodate fillets. Another reason for unequal wall thicknesses could be the need to have sharp internal edges (Figure 4.20). In order to create these sharp edges, a hollow molding is created and results not only in a sharp edge but also unequal wall thicknesses.

In extrusion, the material flow speed through the die needs to be equal at all points of the cross section. There are a number of factors that influence the material flow speed through the die most important of which is material selection. Low alloyed materials are easier to extrude than high alloyed materials. Heat generation during the extrusion process can generate cracks and the complexity of the cross section also has an effect. Complex cross sections demand thicker walls. The main factors that influence wall thickness are extrusion force, extrusion speed, alloy selection, profile shape, as well as the desired surface finish and tolerance specifications.

Compared to steel, aluminum extrusions can be made with more complexity due to aluminum's mechanical properties. Fins for heat transfer, complex pedal (gas or brake) design for strength, or other designs to improve assembly, etc., can be added to extrusions when using aluminum as opposed to sheet metal-forming steel. Fins are one of the most important advantages for aluminum extrusions due to their wide applicability in heat transfer operations as well as adding stiffness to a part. By increasing a part's effective surface area, more heat transfer is possible. By adding fins to an ignition control module or engine control module, the heat produced by the electronics can be dissipated safely to avoid damage to these critical components.

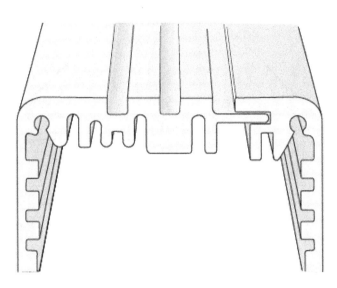

FIGURE 4.21
Example of joining two open sections [17].

4.3.3.3 Hollow Profiles for Joining Extrusions

Extrusions can be joined a number of ways. Many different welding techniques can be employed as well as other creative solutions. The types of welding can include metal inert gas (MIG), tungsten inert gas (TIG), laser, spot, and friction stir welding. Some of the other common joining technologies include blind rivets, self-piercing rivets, compression fit joint, clinching, bolting, and adhesive bonding. There are many advantages by joining several smaller extrusions to a larger part including:

- Easier part handling
- Pressing, surface treatment, and machining which can be done on a more rational basis
- Smaller extrusions which can be produced with less material thickness, better accuracy, and lower die costs

Joints in aluminum extrusions can be sophisticatedly hidden by making it a part of a fluted design. Using one of the joining techniques mentioned above, the two sections can be mated as shown in Figure 4.21. Extrusions can also be made to have screw ports designed integrally with the part. This allows for the use of self-taping screws to join the parts together. These screw ports can be closed or open to some degree. Closed screw ports are typically used when the joint requires a more robust fastener (Figure 4.22).

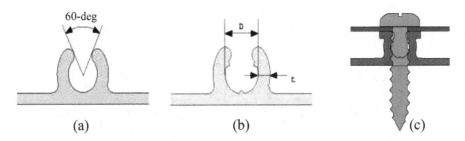

FIGURE 4.22
(a, b) closed screw ports (c) joining two hollow sections [17].

4.3.3.4 Design Guidelines For Extrusions

When designing extrusions, there are some general guidelines that help reduce cost and make the extrusions a more suitable end-product. These are as follows:

1. Avoiding sharp edges and tips is a good standard to follow when designing any part. It reduces stress concentrations in addition to making the profiles easier to extrude (Figure 4.23 (a)).

2. Solid profiles are also preferred for extrusion processes due to them being easier to manufacture (Figure 4.23 (b)).

3. Generally reducing the number of cavities reduces production costs, increases die stability. The increased die stability makes the part quality better and increases die longevity (Figure 4.24).

4. Common practice is to keep the width-to-height ratio to approximately 1:3 to ensure the strength of the die is not compromised. Die strength is an important consideration when looking at the life of an extrusion die.

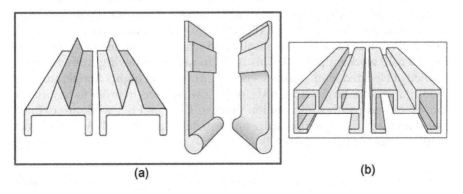

FIGURE 4.23
(a) Sharp tips; (b) Integrated solid profile [17].

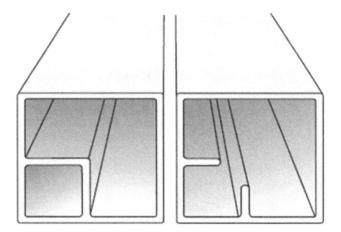

FIGURE 4.24
Section on the right is more suitable than on the left [17].

4.3.4 Drawing

Another bulk deformation process similar to extrusion is called drawing. Drawing is a metalworking process which uses tensile forces to stretch metal. As the metal is drawn (or pulled), its cross section becomes thinner into the desired shape and thickness. Drawing is classified into the following types:

- Sheet metal drawing
- Wire drawing
- Bar drawing
- Tube drawing

This section will touch on bar and tube drawing where sheet metal drawing is covered later in this chapter. For wire drawing, bar drawing, and tube drawing, the cylindrical stock material is drawn through a die to reduce its diameter and increase its length (Figure 4.25). Bar drawing concept is similar to wire drawing. Drawing is typically performed at room temperature, making it a cold working process. This is not to say that drawing cannot be performed at elevated temperatures. Usually, drawing at higher temperatures is required to work large wires, rods, or hollow sections in order to reduce the force required for the deformation process. Cold and hot work are defined based on the working temperature of the material and its melting point. Cold work is defined as (Equation 4.2):

$$\frac{T}{T_m} < 0.3 \tag{4.2}$$

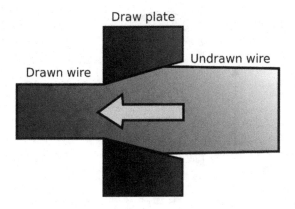

FIGURE 4.25
Schematic of drawing manufacturing process [18].

Warm work is defined as (Equation 4.3):

$$0.3 < \frac{T}{T_m} < 0.6 \tag{4.3}$$

Hot work is defined as (Equation 4.4):

$$\frac{T}{T_m} > 0.6 \tag{4.4}$$

Where T is the working temperature of the material and T_m is the melting point of the material. Aluminum is a growing material in place of copper for the drawing cables due to aluminum's lower cost and weight.

4.3.5 Rolling

Rolling is a bulk deformation process where the raw material thickness is reduced due to compressive forces imposed by two or more opposing rollers. These rotating rolls perform two main functions:

- Pull the workpiece into the space between the rollers using the friction force between the workpiece and the rollers.
- Squeeze the workpiece into the final cross section.

Rolling is classified into two types: by the geometry of the work and by the temperature of the work. Flat rolling and shape rolling are classified into the geometry category, and hot rolling and cold rolling are classified into the temperature category. Hot rolling is where the raw material is heated to ease

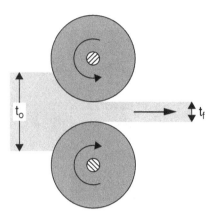

FIGURE 4.26
Schematic of the geometry of the flat rolling process [19].

deformation and is a common rolling process when large deformations are required. Cold rolling is where the workpiece is rolled under cold conditions described using the equations in Section 4.3.4.

Flat rolling is used to reduce the thickness of a rectangular cross section. Figure 4.26 below shows some of the geometry of a flat rolling operation. Two important terms for flat rolling are draft and reduction. A draft is the amount of thickness change (Equation. 4.5). Reduction is the draft expressed as a fraction of the starting material thickness (Equation. 4.6).

$$d = t_0 - t_f \tag{4.5}$$

$$r = \frac{d}{t_0} \tag{4.6}$$

Here d is the draft, t_0 is the starting thickness, t_f is the final thickness, and r is the reduction.

Shape rolling is where a square cross section is formed into a shape (I-beam, Train rails, L-beams, etc.). The workpiece is deformed into a contoured cross section as opposed to a flat cross section (like in flat rolling). This process is performed by passing the original material through rollers that have the opposite contour than what is desired. One type of specialty shape rolling is called thread rolling. Thread rolling is another bulk deformation process that is used to form threads on cylindrical parts. Figure 4.27 shows the schematic of a flat die thread-rolling process. This process is used for the mass production of screws and bolts because typical production rates are around one piece per second. Forming and rolling produce no swarf and

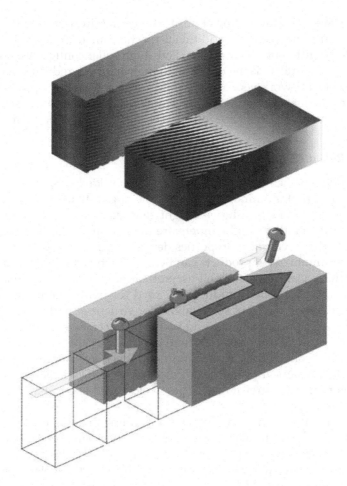

FIGURE 4.27
Schematic of the thread-rolling process [20].

less material is required because the blank size starts smaller than a blank required for cutting threads. Therefore, there are typically 15 to 20% material savings in the blank, by weight. The thread-rolling process is preferred over thread machining because it yields higher production rates, more conservative material consideration, stronger work-hardened threads, and better fatigue resistance due to a compressive force exerted by the rollers. Also, the stress concentration effects are minimized in thread rolling.

Among the different types of rolling processes, there are a number of different rolling configurations or mills [14]. Two-high roller configuration is typically used for initial thickness reduction of original material. Three-high roller configuration is where the workpiece is redirected between a different pair of rollers that move in the opposite direction. This is a complicated configuration that needs some additional mechanism to pass the workpiece between the different sets of rollers. Four-high roller configuration is where a smaller set of rollers is in contact with the workpiece reducing the contact area between the workpiece and the rollers. This means less force is required to deform the material; however, the smaller rollers are more susceptible to deformation and require large rollers to be stacked on top of them.

4.3.5.1 Rolled Aluminum Products

Rolled products including plate, sheet, foil, or welded tubes are the second largest fraction of aluminum in automobile applications. Using aluminum for these applications reduces weight and enhances part performance. Several different alloys and tempers are utilized and have been developed to provide the mechanical properties desired for a variety of parts and use cases. Aluminum sheet products can be manufactured with special surface topographies, claddings, as well as pretreatments for lubrication, joining, and painting. Hot rolled products are used for many automotive applications, particularly the 5xxx series alloys for structural parts, such as wheel stock, suspension components, and body reinforcements. Cold rolled products are typically sheet or plate, where the final thickness (or gauge) is processed using cold rolling. Annealing treatments are commonly needed to adjust the mechanical properties to meet the specifications of the original equipment manufacturer (OEM). These cold rolled sheets have tight tolerances (Table 4.3).

Clad material is fabricated by rolling a cladding material onto one or both sides of core material. The core materials are typically 3xxx or 6xxx series alloys where 4xxx series alloys are chosen for cladding. The purpose of cladding is to increase a material's strength, brazeability, surface properties, corrosion resistance, etc. Many of the rolled products for automotive applications have a variety of surface conditions (pretreatments, precoatings, etc.). These pretreatments or precoatings can include lubricants (oil), dry film lubricants (hot melts), or something like a Ti/Zr-conversation layer. The reasoning behind pretreatments is for the purpose of surface protection during

TABLE 4.3

Typical Tolerance Limits for Cold Rolled Sheets

Rolling Width (mm)	Thickness Tolerance (mm)
Up to 1000	± 0.05
1000–1250	± 0.09
1250–1600	± 0.10
1600–2000	± 0.12
2000–2500	± 0.14

the handling and transport phase, the improvement of tribological conditions during the forming process, and for the preparation of surface joining processes.

4.3.5.2 Special Alloys and Tempers

Different alloys and tempers are used to meet the quality requirements of numerous automotive parts. A combination of strain hardening and partial annealing is used to produce the H28, H26, H24, and H22 series of tempers. The materials are strain hardened more than is specified in order to achieve the required mechanical properties. After the strain hardening, the materials are reduced in strength by partial annealing. Fully annealing returns materials that have been cold worked to the O temper (a soft ductile condition). However, some materials cannot be heat-treated. These can be seen in Table 4.4.

- 1xxx series: Al 99.5
- 3xxx series: Al–Mn
- 5xxx series: Al–Mg (Mn)

There are a number of basic codes to identify the temper, heat treatment, and strain-hardening measures used on aluminum as listed in Tables 4.4 to 4.7.

The two main body sheet alloy groups are EN 6xxx Al-Mg-Si and EN 5xxx Al-Mg-Mn. The 6xxx series alloy has good formability (T4) age hardenable (T6), good surface finish and corrosion resistance, and requires solution annealing above 500 °C. The 5xxx series alloy has good formability (O), good corrosion resistance for less than 3% Mg (for greater than 3% Mg, prone to intercrystalline corrosion), and good strain hardening.

TABLE 4.4

Basic Aluminum Heat Treatment Designations

Code	Description
F	**As fabricated:** No special control has been performed to the heat treatment or strain hardening after the shaping process
O	**Annealed:** This is the lowest strength and highest ductility temper
H	**Strain hardened:** Applied to wrought produced only. The designation is followed by two or more numbers
W	**Solution heat-treated:** This is seldom encountered because it is an unstable temper that applies only to alloys that spontaneously age at ambient temperature after heat treatment
T	**Solution heat-treated:** Used for products that have been strengthened by heat treatment, with or without subsequent strain hardening. The designation is followed by one or more numbers

TABLE 4.5

Heat Treatment Temper Codes

Code	Description
T1	Cooled from an elevated temperature-shaping process and naturally aged to a subsequently stable condition
T2	Same as **T1**. Additionally, it is cold worked
T3	Solution heat-treated, cold worked, and naturally aged to a subsequently stable condition
T4	Same as **T3** but without the cold work
T5	Same as **T1** but artificially aged
T6	Solution heat-treated the artificially aged
T7	Solution heat-treated the overaged/stabilized
T8	Same as **T6** but cold worked before aging
T9	Same as **T8** followed by cold work
T10	Cooled from an elevated temperature-shaping process, cold worked, then artificially aged

4.3.6 Sheet Metal Forming

Sheet metal forming is a wide subject that has many intricacies that require an indepth knowledge. Only an overview will be covered in this section. There are several types of sheet metal-forming processes, such as drawing, rolling, hydroforming, stretch forming, explosive forming, plastic forming, stamping, etc., that are covered in standard manufacturing technology textbooks [21, 22]. Drawing, rolling and hydroforming have either been discussed previously or will be covered briefly in a later section. This section focuses on stamping. Stamping is a process that involves pressing sheet metal into a die to form an infinite number of shapes. Many body parts in an automobile are made this way. Stamping is a process where every second counts. Even a lag time of 0.1 of a second can be considerably costly over many years of operation.

There are numerous variables that affect the stamping process. The most important parameters include blank size, blank shape, component shape, press tonnage, punch speed, and friction coefficient between blank and

TABLE 4.6

Strain Hardening Codes

Code	Description
H1	Strain hardened only
H2	Strain hardened and partially annealed
H3	Strain hardened and stabilized
H4	Strain hardened and lacquered or painted

TABLE 4.7

Summary of Temper Designations

Code	Description
F	As fabricated. No control over the amount of strain hardening; no mechanical property limits
O	Annealed, recrystallized. Temper with the lowest strength and greatest ductility
H1	Strain hardened. H12, H14, H16, and H18. The degree of strain hardening is indicated by the second digit and varies from quarter hard (H12) to full hard (H18), which is produced with approximately 75% reduction in area
H2	Strain hardened and partially annealed. H22, H24, H26, and H28. Tempers ranging from quarter hard to full hard obtained by partial annealing of cold worked materials with strengths initially greater than desired
H3	Strain hardened and stabilized. H32, H34, H36, and H38. Tempers for age-softening aluminum–magnesium alloys that are strain hardened and then heated at a low temperature to increase ductility and stabilize mechanical properties
H112	Strain hardened during fabrication. No special control over the amount of strain hardening but requires mechanical testing and meets minimum mechanical properties
H321	Strain hardened during fabrication. The amount of strain hardening is controlled during hot and cold working
H323, H343	Special strain hardened, corrosion-resistant tempers for aluminum–magnesium alloys

tooling, etc. If one or more of these parameters is ignored or forgotten about, defects in the stamped components can occur. The four most common aluminum grades available as sheet metal are 1100-H14, 3003-H14, 5052-H32, and 6061-T6. Some common stamp defects include earing, tearing, wrinkling, loose metal, spring back, etc. Figure 4.28 shows typical drawing failures of a cylindrical cup [23]. These defects result in a loss of time and create considerable scrap.

4.3.7 Hydroforming

Hydroforming is one of the sheet- or tube-forming processes where the material positioned inside the mold is filled with liquid water or air, and very high pressure is given from both ends of the material in order to deform the material as the mold moves. Tube hydroforming is generally defined as high or low pressure with the demarcation point of around 83MPa. There are two main types of hydroforming: tube hydroforming and sheet hydroforming. Tube hydroforming is used when a complex cross section shape is needed. In tube hydroforming, a section of cold rolled steel or aluminum tubing is placed in a closed die set where a pressurized fluid is forced into the ends of the tube and the tube reshapes within the confines of the mold.

Eared cup   Punch nose radius failure

Die entry radius failure   Gridded blank to visualize strain

FIGURE 4.28
Different types of failure of cylindrical cup drawing [23].

Figure 4.29 shows the working principle of tube-hydroforming simulation using AutoForm [24]. The figure shows the four steps of the process. Figure 4.30 shows a sheet that can be formed by exerting fluid pressure into the die cavity [25, 26]. Figure 4.30 shows the initial step (top die in the open position), and the final step after the sheet is formed by water or air under pressure (with the top die in closed position). This process produces better-quality drawn parts compared to mechanical stamping, for example.

Figure 4.31 shows two examples of tubes produced by the hydroforming process [25].

A closed section material is hydroformed to produce a variety of cross sections and shapes along the tube length offering complex shapes for assembly advantages, lighter components, and varying stiffness when desired. Hydroforming is being used to replace stamped and welded parts with hydroformed parts, such as crossbeams, roof beams, as well as B-pillars. In

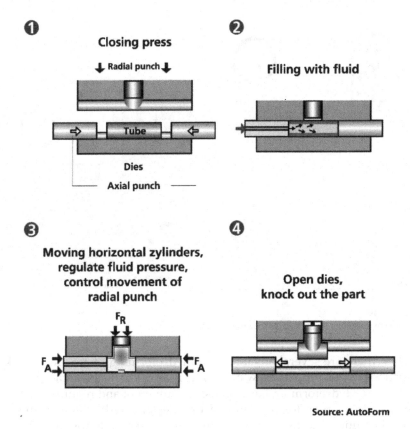

FIGURE 4.29
Working principle of tube-hydroforming simulation using AutoForm [24].

addition, there are more automotive applications for hydroforming listed below.

- Body shell
- Driveshaft
- Assembled camshaft
- Exhaust systems
- Engine cooling system
- Radiator frame
- Engine bearer
- Frame structure
- Axle elements

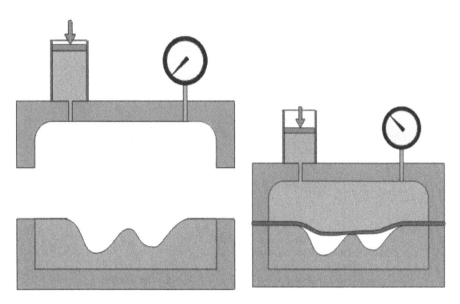

FIGURE 4.30
Working principle of sheet hydroforming process [26].

While there are several advantages to hydroforming, there are some key limitations of this process. Slow cycle time is a critical disadvantage for mass production. Hydroforming has expensive equipment and requires an extensive knowledge base for process and tool design. Additionally, it can create challenging welding techniques in order to assemble the hydroformed

FIGURE 4.31
Examples of tube hydroforming [27].

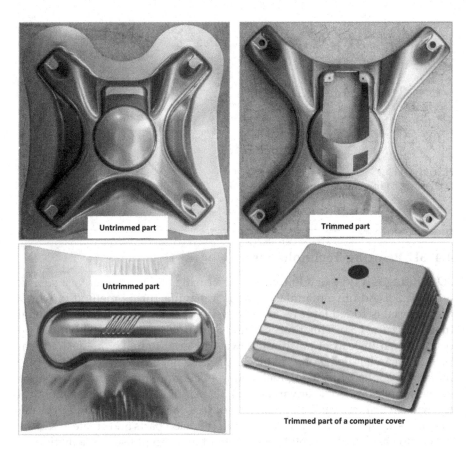

FIGURE 4.32
Examples of sheet steel components manufactured by fluid forming process [Courtesy of FluidForming Americas, LLC, Ref. 28]

components. Lastly, the hydroforming process requires more physical space for storing and recalculating the working fluid (water).

Fluid forming [FluidForming Americas, LLC, Ref. 28] is also a hydro-forming process, which is bladder-free metal forming technology that offers unprecedented design flexibility at lower maintenance costs. In this, every step of the forming process from blank design to material flow to clamping pressure to volume and water pressure is variable and controlled with minimal or no springback, FluidForming achieves finer, and more detailed results. Forming pressures of up 4000 bar (60,000 psi) yield better (Six Sigma) accuracy. This process has also 99.996% first-pass-yield rate. Figure 4.32 shows examples of steel stamped parts manufactured by fluid forming process [28].

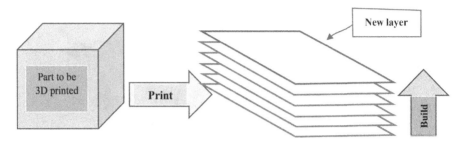

FIGURE 4.33
Principle of additive manufacturing.

4.4 3D Printing Technologies

3D printing is a type of additive manufacturing and is another vast subject area [29, 30]. Additive manufacturing is a manufacturing method in which the material is added layer by layer by using data from a three-dimensional model (Figure 4.33). This process builds the final part from the bottom up or top-down instead of machining away material to make the desired features. Research is being done in the 3D printing field in order to manufacture industrial applications using metals like steel or aluminum instead of resin or thermoplastics. Some advantages of 3D printing are that it is economical to own, 3D printers can be operated in offices, labs, home, etc., and printing machines accept digital inputs from solid modeled parts and creates solid 3D parts and easily makes prototypes for proof on concept uses. Schematic representation of the 3D printing technique, also known as fused filament fabrication is shown in Figure 4.34. In this process, a filament (shown as a) in the figure) of plastic material is fed through a heated moving head b) that melts and extrudes it depositing it, layer after layer, in the desired shape (shown at position c) in the fiure). A moving platform e) lowers after each layer is deposited. For this kind of technology additional vertical support structures d) are needed to sustain overhanging parts.

A schematic representation of stereolithography is shown in Figure 4.35, in which a light-emitting device (laser or DLP, shown at position a) in the figure) selectively illuminates the transparent bottom (shown at c) in the figure) of a tank (shown at position b) in the figure) filled with a liquid photopolymerizing resin; the solidified resin (shown at position d) in the figure) is progressively dragged up by a lifting platform (shown at position e) in the figure).

As shown in Figure 4.34, stereolithography apparatus' position a platform just below the surface of a vat of liquid UV resin. This liquid UV resin is cured when UV light of certain wavelengths passes through the liquid and polymerization occurs. A UV laser beam traces the first slice of the desired

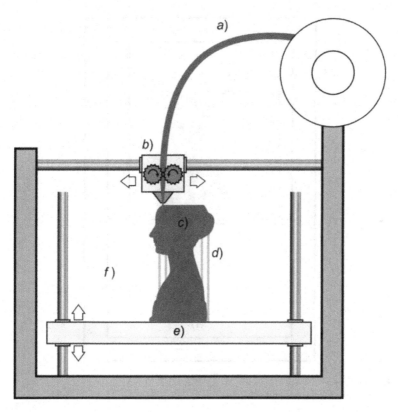

FIGURE 4.34
Schematic representation of the 3D printing technique (FFF) [29].

object on the surface of this liquid causing a very thin layer of photopolymer to harden. The platform is then lowered very slightly and another slice is traced out and hardened by the laser. This process of tracing, hardening, and lowering the platform is repeated until a complete object has been printed and can be removed from the vat of UV resin. Once the object is removed, it is drained of any excess resin and fully cured.

Fused deposition modeling (FDM) is where an extruder head moves in two principal directions over a table (Figure 4.36). The thermoplastic filament is extruded through a heated die and out of the extruder head as the extruder follows a predetermined path. Each layer is roughly 125–325 μm depending on the amount of fidelity desired. Common materials for this method include acrylonitrile butadiene styrene (ABS), polyamide, polycarbonate, polyethylene, and polypropylene. In some special applications, materials, such as silicon nitrate, PZT, aluminum oxide, hydroxyapatite, and stainless steel, can be used for a variety of structural, electroceramic, and bioceramic applications.

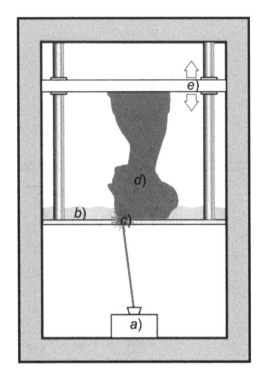

FIGURE 4.35
Stereolithography schematic [30].

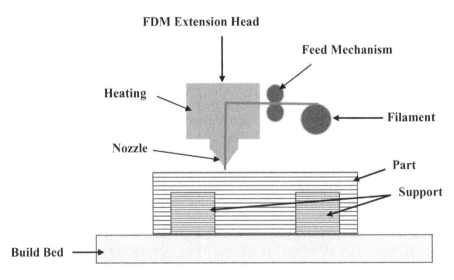

FIGURE 4.36
Schematic of fused deposition modeling (FDM) (adapted from [31]).

Other 3D printing technologies listed below are available in the literature but not discussed here in detail.

- Selective laser sintering (SLS)
- Multijet molding
- Additive laser manufacturing (ALM)
- V-flash printer
- Desktop factory
- Fab@Home

One of the 3D printing companies, Azoth LLC focused on the production of blueprint parts made on demand through additive manufacturing (AM) [32]. Their primary goal has been to transition physical inventory (indirect materials) in manufacturing facilities into digital inventory 3D-printed on demand. The technology and quality systems developed to fulfil this vision resulted in unique manufacturing capabilities that have allowed Azoth to expand its offering to end-use products in market sectors, such as medical, consumer goods, and automotive.

One of the major advantages of additive manufacturing is the ability to create a lightweight structure. A few of the ways that this can be achieved is through the use of manual design, computer-driven design, and the basic lattice packages that can be used in aiding design. Figure 4.37 displays a component that has been cut in half to display the internal geometry. The part was manufactured via metal FDM on the Desktop Metal's Studio System. This part was 40% lighter by using a standard honeycomb option than the

FIGURE 4.37
3D-printed component via metal FDM process (courtesy of Azoth [32]).

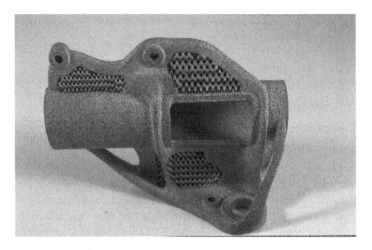

FIGURE 4.38
Computer-driven 3D-printed component showing lattice structures for lightweight using topology optimization (courtesy of Azoth [32]).

original part. Figure 4.38 also represents a computer-driven design. As can be observed, the lattice structures were added for lightweighting and then topology optimization was used to simulate the loads the part would experience.

In turn, the material was atomically removed through the program to optimize the strength-to-weight ratio, resulting in a fully optimized geometry that can only be manufactured through 3D printing. With the binder jetting technology used in 17-4 PH, a 28% weight savings was generated.

Figure 4.39 shows a gear design in which the designer has manually removed material based on his/her application knowledge. This has increased the part complexity, but not cost with 3D printing. These parts were manufactured with metal binder jetting (MBJ) in 17-4 PH stainless steel.

FIGURE 4.39
Gears manufactured using metal binder jetting (MBJ) (courtesy of Azoth, LLC [32]).

FIGURE 4.40
3D parts manufactured via metal binder jetting (courtesy of Azoth, LLC [31]).

Figure 4.40 shows the other options of adding lattice structures to a part which cannot be machined using the traditional methods. For getting a better visual effect, these were produced via binder jetting in 316L. In addition to these, Azoth manufactures 3D-printed parts using several other methods [32].

With a variety of current technologies, 3D printing has applications in many different fields. For example, medical apparatuses can give custom-fit prosthetics and braces to patients quickly. Figure 4.41 shows an example of the assembly drawing of two designs of a prosthetic foot designed by Kettering University students ("Kettering Prosthetic Foot") as a part of their capstone class project that the author supervised [33]. They used Siemens NX software for the CAD, motion simulation, and the structural finite element analysis. The students performed several design iterations of their conceptual ideas before finalizing their design with split foot shown on the right of Figure 4.41. Figure 4.42 shows two prototypes of the same prosthetic split foot designs printed by using the SLS method at Azoth, LLC.

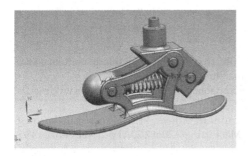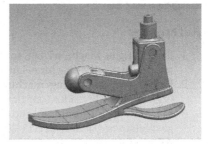

FIGURE 4.41
Two design iterations of the prosthetic foot assembly using Siemens NX software showing solid foot (left) and split foot (right) [33].

FIGURE 4.42
Two design iterations of the prosthetic foot assembly using Siemens NX software (courtesy of Azoth, LLC).

Many industrial applications of 3D printing technology in the automotive industry and other advanced manufacturing are also growing. Airbus made plans to make a 3D printer that is large enough to make planes from the ground-up [34–36]. Additionally, NASA has considered 3D printing in space to enabling astronauts to create their own equipment during the trips. Other useful references on machinability ratings of aluminum and a list of different applications of aluminum are given in references 37 and 38.

References

1. James. "Lathe Machine Diagram." Colturn company, available at: colturn.co.za/my-product/lathe-machine/lathe-machine-diagram/
2. Frobes vertical milling machine available at: "https://commons.wikimedia.org/wiki/File:Fresadora_vertical.JPG, March 6, 2008.
3. Machining of Aluminum and Aluminum Alloys: https://materialsdata.nist.gov/bitstream/handle/11115/200/Machining%20of%20Al.pdf?sequence=3&isAllowed=y
4. *Machinery's Handbook*, 30th Edition, by Erik Oberg [Digital, 2016..
5. Typical hand ground cutting tool angles for the lathe: https://en.wikipedia.org/wiki/Tool_bit#/media/File:Tool_Bit_Geometry.JPG

6. Geometry of drill bits available at: https://www.engineeringenotes.com/industrial-engineering/drilling-machine/15-main-parts-of-twist-drill-machine-tools-industrial-engineering/22826

7. Images of drill bits available at: https://upload.wikimedia.org/wikipedia/commons/c/c3/Drillbits.jpg

8. Die casting process available at: https://en.wikipedia.org/wiki/Die_casting

9. Permanent mold casting process available at: https://en.wikipedia.org/wiki/Permanent_mold_casting

10. Die casting/permanent mold casting: https://en.wikipedia.org/wiki/Gear; https://en.wikipedia.org/wiki/Sprocket; https://en.wikipedia.org/wiki/Gear_housing; https://en.wikipedia.org/wiki/Piston

11. Sand casting process available at: https://en.wikipedia.org/wiki/Sand_casting#/media/File:Haandform-e.png

12. Various forging process available at: https://en.wikipedia.org/wiki/Forging

13. Plots of forces in extrusion available at: https://commons.wikimedia.org/wiki/File:Extrusion_force_plot.png

14. (a) Aluminum Automotive Manual, 2015: https://www.european-aluminium.eu/resource-hub/aluminium-automotive-manual/; (b) https://www.aluminum.org/resources/industry-standards/aluminum-alloys-101

15. Crash Box CAD Drawing by Paris Velikis available at: GrabCAD.com

16. Extruded aluminum parts with several hollow cavities: https://en.wikipedia.org/wiki/Extrusion#/media/File:Extruded_aluminium_section_x3.jpg

17. Aluminum Extrusions: https://www.aec.org/page/extrusion-applications-transportation

18. Schematic of wire drawing process: https://en.wikipedia.org/wiki/Wire_drawing#/media/File:Wiredrawing.svg

19. Schematic of rolling process: https://en.wikipedia.org/wiki/Rolling_(metalworking)#/media/File:Laminage_schema_gene.svg

20. Schematic of a flat die thread rolling process: https://en.wikipedia.org/wiki/Threading_(manufacturing)#/media/File:Screw_(bolt)_14-n.PNG

21. *Fundamentals of Manufacturing Engineering and Technology*, 7th Edition, by S. Kalpakjian and S. Schmid, Pearson Education, 2014.

22. *The Science and Engineering of Materials*, 4th Edition, by Donald R. Askeland and Pradeep P. Phule, Chapter 6, Brooks/Cole publisher, 2003.

23. Virtual and Real Forming of Sheet Metal – A Classroom Scenario by Raghu Echempati, William J. Riffe, K. Joel Berry, Proceedings of ASEE 2000 Annual Conference.

24. Principle of Tube Hydroforming by AutoForm: https://www.autoform.com/en/glossary/hydroforming/

25. Sheet hydroforming: https://en.wikipedia.org/wiki/Hydroforming#/media/File:Idroformatura_animata.gif

26. What is hydroforming: https://americanhydroformers.com/what-is-hydroforming/; https://www.ffamericas.com/

27. Example of hydroformed tubes: http://www.h-htube.com/custom-hydroforming.html

28. Example of fluid formed parts: ffamericas.com

29. Paolo Cignoni, Schematic representation of Fused Filament Fabrication: https://en.wikipedia.org/wiki/3D_printing#/media/File:Schematic_representation_of_Fused_Filament_Fabrication_01.png

30. Paolo Cignoni, Schematic of stereo lithography: https://en.wikipedia.org/wiki/3D_printing#/media/File:Schematic_representation_of_Stereolithography.png

31. Alafaghani, A."Experimental Optimization of Fused Deposition Modelling Processing Parameters: A Design-for-Manufacturing Approach,", et al, *Procedia Manufacturing* 10 2017: 791–803, Elsevier.

32. 3D Printing technologies available at: http://www.azoth3d.com/

33. Unpublished capstone final project on "Kettering Prosthetic Foot", Summer 2020 group of students, Kettering University, Flint, MI, 2020.

34. Geometry, weights and topology by APWORKS, available at: https://apworks.de/en/design/; and https://apworks.de/en/scalmalloy/#application

35. Design and topology optimization software information available at: https://www.altair.com/structures-applications/

36. Weight watchers by Develop 3D, available at: https://develop3d.com/product-design/weight-watchers-light-rider-3d-printing-topology-optimisation-design/

37. Skill Builder: https://pl.quakerchem.com/wp-content/uploads/pdf/skill_builders/no10_machineability_ratings.pdf

38. Drive aluminum has many resources available at: https://www.drivealuminum.org/

5

Car Body Structures

5.1 Body Design Concepts and Crash Performance

The core element of any car is the body structure. The body is an extremely important part of any automobile for a number of reasons and demands key design considerations. The body should connect all the components, house the drive train, as well as provide crash load paths. Additionally, the body structure should be able to carry and protect passengers as well as cargo. Lastly, the body structure should be joinable with other components in addition to being lightweight to provide optimum fuel economy and performance. The following are the three main design concepts. However, tubular space frame design and backbone chassis design are discussed.

1. Body-on-frame (Figure 5.1)
2. Monocoque (or single shell) (Figure 5.2)
3. Unibody (Figure 5.3)

Body-on-frame is a design concept [1] where the body structure is completely separate from the frame. This is the oldest structural vehicle design. The frame typically consists of two connected rails where the suspension and power train are housed and connected. This type of concept is also referred to as ladder frame construction since the frame (from a top view) looks akin to a ladder.

There are some distinct advantages as well as disadvantages to the body-on-frame design. The body-on-frame concept is easier to design, build, and modify, which has historically been a major factor. However, since CAD is commonplace, it is less of an issue for modern vehicles but still an advantage for coach-built vehicles. The ladder frame design is more suited for heavy-duty usage and can be more durable than other concepts. In addition to being more durable, the body-on-frame design is easier to repair after accidents and provides a better ride quality for SUVs. Despite the advantages, there are some important disadvantages to consider. In the body-on-frame design, only the longitudinal members bear the longitudinal forces caused by acceleration

FIGURE 5.1
Body-on-frame design concept [1].

and braking. Additionally, only the lateral and cross members provide resistance to lateral forces and this increases the torsional rigidity. Those two disadvantages compromise safety because the rails do not deform under impact and transmit more impact energy into the cabin. The frame design is a two-dimensional structure and this lends itself to having poor torsional body stiffness. Lastly, the body-on-frame design tends to take up a lot of valuable space. This forces the center of gravity to go up causing the vehicle to be more prone to rollover crashes despite the frame itself being very heavy.

The monocoque design [2] utilizes an external skin to support some or most of the load. This contrasts the body-on-frame design (Figure 5.2 without the door panels). The design integrates the body and chassis together. In this design, the floor pan serves as the main structural element to which all the mechanical components are attached.

Just like every design, the monocoque design has its advantages and limitations. The monocoque design structure has good crash protection since crumple zones can be built right into the structure. The monocoque structure defines the overall shape of the car and incorporates the chassis right into the body. This requires less fastening and, consequently, less failure points. The monocoque design is space-efficient and lends itself well to mass production using aluminum. Despite the advantages discussed above, the monocoque design requires high tooling costs that hinder its application for small scale production. Additionally, the pure monocoque structure can be relatively heavy but more recently, the design has lost weight due to the use of aluminum instead of

FIGURE 5.2
Monocoque design concept [2].

FIGURE 5.3
Unibody design concept [3].

steel. Lastly, the rigidity-to-weight ratio is fairly low as the design is intended to benefit space efficiency rather than strength. The pressed sheet panels are not as stiff as structures made from tubes or other closed sections.

The unibody design concept [3] uses a system of box sections, bulkheads, and tubes to provide strength to the vehicle (Figure 5.3). Unibody is a design technique that the body is integrated into a single unit with the chassis rather than have the body on frame. Unibody design is similar to monocoque design. Most modern automobiles are not pure monocoques; rather, they are unibody construction. In some cases, unibody and monocoque are interchangeable but according to a few views, there are slight differences. In monocoque construction, the outer shell (Class A surface) is actually part of the body structure whereas, in unibody construction, the Class A surface is not.

The unibody design concept allows for a significant weight reduction of the car body and enables a more compact, yet spacious, vehicle configuration. Since the deformation zones that absorb impact energy can be engineered right into the body, crash safety is enhanced. The design allows for suspension components as well as the power train to be directly mounted to the unibody. Lastly, other important components, such as the subframe and firewall, can be integrated into the unibody design. Despite the level of integration, the rigidity of the car body is not as optimal as the monocoque design. In the event of a crash, the unibody design is very expensive to repair. Lastly, the unibody design is not as sophisticated as the monocoque design since the windshield and rear window glass make the monocoque design stronger.

Lastly, the tubular space frame and backbone chassis design concepts [4, 5] are used in niche applications and not for mass production. The tubular space frame design is typically used in sports cars (Figure 5.4). This design

FIGURE 5.4
Jaguar C-type tubular space frame design [4].

FIGURE 5.5
Backbone chassis of 1923 Tatra 11 car [5].

employs dozens of tubes or other rod-shaped components that are welded together in different directions in order to provide the required mechanical strength against forces from anywhere. This design is believed to enhance the rigidity-to-weight ratio of the vehicle.

The backbone chassis design concept is very simple in construction (Figure 5.5). It consists of a strong tubular backbone which connects the front and rear axles, which provides nearly all the mechanical strength. The entire drive train, engine, and suspension components are connected to both ends of the backbone. The modular design of this system enables various configurations and the more axles with all-wheel drive, the more the cost benefit increases. Despite these benefits, the manufacturing of the backbone chassis is complicated and expensive. The design is strong enough for small sports cars but is not suitable for larger high-performance sports cars. Lastly, this type of design offers absolutely no protection from side impacts and offset crashes.

One of the most important functions of a car body structure is the performance in the event of a collision. The body structures should allow the crash energy to be absorbed into the car body by dissipating the energy among the structural members and not the occupants of the vehicle. Engineers must design load paths in order to protect the passengers. This is a critical function of the car body structure and the use of different materials and designs can affect crash performance.

There are three main types of crash scenarios: frontal impact, side impact, and rollover events. Each scenario must be considered for peak crash performance. There are a number of ways that engineers can enhance a vehicle's crash performance. Designing a car that has a low center of gravity reduces

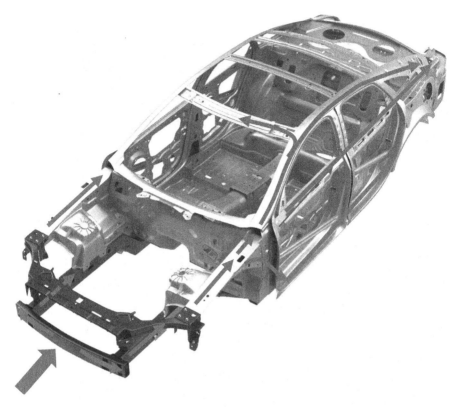

FIGURE 5.6
Collision path for frontal crash (adapted from ref [2]).

the potential for rollover incidents as well as giving the best possible distribution of load paths around the passengers. For a frontal collision scenario, a continuous side member that extends from the front cross-member to the side skirt directs the impact energy into the extremely rigid door sill structure, thus protecting the occupants during a frontal impact (Figure 5.6). Using strong door frames and door structures protects from side impacts (Figure 5.7). Engineering paths for the load to travel transversely through the car can be tricky but it achieved by directing it through the main cross members in the vehicle as shown in these figures.

5.2 Body Design with Aluminum

Using aluminum in automotive design has been increasingly popular. Aluminum body concepts are becoming more common due to technical, economical, and fabrication flexibilities. Aluminum offers a wide variety of

FIGURE 5.7
Collision path for side crash (adapted from ref [2]).

design options due to its low density, improved crash safety, and increased fuel economy (due to its light weight). Due to these properties, aluminum offers the substitution of steel in body structures. Several automotive manufacturers have turned to aluminum to be used in the car body structure; however, none has used aluminum exclusively yet. The following are some examples of US and European cars that contain aluminum in the body structure. An example of this is shown in Figure 5.8

- Ford, GM, Tesla, and Fiat Chrysler
- Audi
- Mercedes E- and S-Class vehicles
- BMW 5- and 7-Series
- Peugeot 307 and 607
- Renault Laguna
- VW Lupo Eco Version

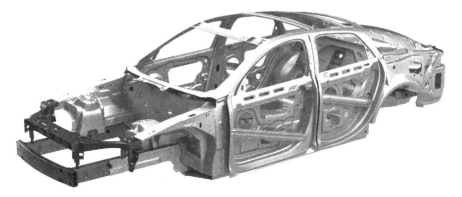

FIGURE 5.8
Carcass of sedan body in white with aluminum engine bay [8].

- Citroen C5
- Volvo V70, S60, and S80
- Land Rover Discover
- Range Rover

European performance vehicles, such as Audi, Mercedes, Jaguar, and BMW, have begun to use aluminum extensively in order to boost performance. For example, the BMW 7 e40 eDrive car is made of carbon-fiber-reinforced polymer (CFRP), tensile steel, and aluminum, which results in a lower curb weight and center of gravity while maintaining a 50/50 axle load distribution [6]. One of the main components that enable the high-performance lightweight is the aluminum front axle. Using aluminum for the axle gives a weight saving of 30%, improved responsiveness of the suspension and steering systems, and easier steering control. Additionally, the rigid, diagonal front axle subframe also carries the steering gears, track control arms and push bars, as well as the antiroll bar. The 7-Series also supports aluminum for the front portion of the frame in order to take impact energy and disperse it throughout the rest of the body frame as well as absorb the energy through deformation. Similar to the BMW 7-Series, Jaguar has integrated 6014-T6 aluminum into its body designs in order to reduce the weight, increase safety, as well as increase fuel economy [7, 8].

The second-generation Audi Space Frame vehicle [9, 10] included 60% sheet panels, 22% castings, and 18% extruded sections (by weight). This is similar to what was shown in Figure 5.2. Designed for higher production volumes than the A8, it used important refinements and new developments in the areas of tools, casting, and joining processes. The number of individual parts was significantly reduced compared to the A8 (from 334 parts to 250 parts). This is shown in Table 5.1. The weight of the all-aluminum body of the D3 (body-in-white plus closures) is 277 kg. AlMgSi alloys similar to EN

TABLE 5.1

Part Count and Weight Reduction between Audi D2 and D3 Models
(Raghu Retyped the Data from Various Sources Including Ref 10)

Audi A8	D2 (1994)	D3 (2000)
Sheet Panels	237	168
Extrusions	47	53
Castings	50	29
Total	334	250

AW- 6060 were used for the extruded components. If necessary, the extruded sections were bent on CNC stretching and rolling machines. Where close tolerances (± 0.3 mm) were required, the semi-finished parts were calibrated and shaped by hydroforming. In addition, also mechanical calibration was adopted for the first time for the A8 as a lower-cost technique.

Figure 5.9 shows an example of the unibody structure by one of the OEMs that is utilizing more aluminum in its body structure design (Figure 5.9). Much of the aluminum additions are in the sheet metal areas; however, some vehicles, such as Mercedes, also use die casting to make some of the aluminum parts. For example, the upper firewall in a Mercedes SL is made using die casting. This enables Mercedes to make high complexity parts and integrate them into a lightweight design. Mercedes is also using friction stir welding in order to join aluminum extrusions. The main floor pan is the major component where Mercedes employs friction stir welding. This allows

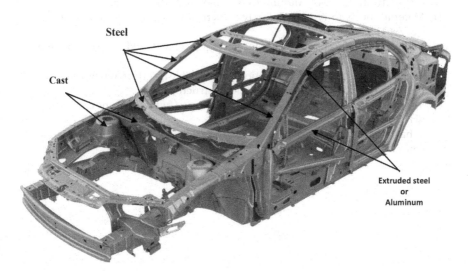

FIGURE 5.9
Material forms in typical unibody structure vehicles [11].

for the car to have very strong, yet lightweight, joints. Additionally, Mercedes uses aluminum for "hang on" parts. These "hang on" parts are usually body components and using aluminum will reduce the weight considerably [10]. Several extrication methods and how they impact the design of the vehicle body structures are discussed in references [12–15].

References

1. Body-On-Frame Design Concept (Author: MotorBlog from Ca, USA): https://upload.wikimedia.org/wikipedia/commons/c/c8/IAA_2013_BMW_i3_%289833675545%29.jpg
2. Monocoque Design Concept: www.shutterstock.com – image # 577964197.
3. Unibody Design Concept: www.shutterstock.com – image # 1579096096.
4. Jaguar C-Type Tubular Space Frame Design: Hmaag, 2 January 2009: https://commons.wikimedia.org/wiki/File:Jaguar_C-Type_Frame.JPG
5. Backbone Chassis of 1923 Tatra 11 car: KapitanT, 2006: https://en.wikipedia.org/wiki/Backbone_chassis#/media/File:ChassisT11.JPG
6. What's your reason not to change? https://www.bmw.com/en/innovation.html
7. Driving performance: https://www.jaguarusa.com/all-models/xe/features/index.html
8. Carcass of sedan body in white with Aluminum Engine Bay: www.shutterstock.com – 577964224.
9. Circular economy: https://www.audi.com/en/company/investor-relations/talking-business/promoting-circular-economy.html
10. Car Body Structures – Aluminum Automotive Manual, 2015.
11. Material forms in typical Unibody Structure vehicles: www.shutterstock.com – 577964196.
12. Vehicle extrication are available at: https://en.wikipedia.org/wiki/Vehicle_extrication
13. Several article on vehicle extrication are available at: http://www.boronextrication.com/2017/05/28/2018-audi-a8-body-structure/
14. Vehicle extrication video available at: https://www.youtube.com/watch?v=4HESUmU7StQ
15. What is vehicle extrication YouTube video available at: https://www.youtube.com/watch?v=v6s4_0PdzrI

6

Joining Technologies

With lightweighting becoming an increasingly popular practice, being able to assemble dissimilar materials, such as plastics, composites, and metals, is becoming a more important issue. Since aluminum is the most common material used in lightweight mechanical applications, joining technologies used for the assembly of aluminum parts for automotive applications is a growing subject. References [1] to [3] cover this subject matter in detail.

Materials joining is based upon three principles:

1. Material coalescence
2. Interlocking joints
3. Frictional connection

For material coalescence, the materials are held together by atomic or molecular binding forces. In this case, the atoms and/or molecules must be placed in close proximity to each other by processes without the presence of heat (solid-state welding or diffusion), processes based on the mixing in a liquid state (fusion welding), or processes with and addition of a third, generally hardening, liquid substance (soldering or adhesive bonding. Interlocking joints are formed by the interlocking of two materials or by the anchoring of additional elements into or inside the corresponding materials (mechanical joints). Frictional Connections are the result of friction between the involved materials, enhanced by the application of an external force (shrinking a hub onto a shaft).

6.1 Aluminum Welding

Welding is defined as the joining of materials by the use of heat and/or force, with or without a filler material. Aluminum welding is widely established and has been developed into an important method of joining. Welding is a commonly used fabrication process that joins pieces of materials. This can be accomplished by partially melting the workpieces, sometimes also adding a filler material, to form a pool of molten material that solidifies, cools, and, subsequently, becomes a strong joint. Other welding methods use pressure with or without heating to produce the weld. Different energy sources are used to produce the weld, including electricity, laser, electron beam, and friction. The main types of welding are:

- Metal arc welding
- Beam welding
- Electric resistance welding
- Brazing
- Solid-state welding

6.1.1 Arc Welding

Arc welding is a fusion welding process which uses a power supply to initiate and maintain an electric arc between an electrode and the base material to create a molten metal pool at the welding point. Arc welding can be accomplished either with direct current (DC) or alternating current (AC). The welding region is usually protected by some type of shielding gas, vapor, or slag.

Arc welding is categorized by the type of electrode, consumable or nonconsumable. For aluminum, metal inert gas (MIG) welding (also referred to as gas metal arc welding), tungsten inert gas (TIG) welding (often referred to as gas tungsten arc welding), and plasma arc welding (PAW) are the most common. MIG welding is a semiautomatic or automatic welding process where a continuously fed consumable wire acts as both the filler material and the electrode (Figure 6.1). The welding wire uncoils automatically from a reel to the welding torch. Heat is produced by an arc between the electrode and the base material. MIG welding is a versatile welding process that is suitable for practically all metals, easy to perform on a wide variety of materials (heavy or light gauge), and requires little to no post-welding finishing processes. Differentiation can be made between MIG welding and metal active gas (MAG) welding. For steel welding, active gas mixtures (mainly argon-based gas mixtures which contain oxygen or carbon dioxide) are preferred. In contrast, for aluminum and most other metals, inert shielding gases (argon, helium, or mixtures of these two) are exclusively utilized.

TIG welding is often referred to as gas tungsten arc welding (GTAW) and was developed earlier than MIG welding. TIG welding is still widely used to join aluminum alloy products (Figure 6.2). The small intense arc provided by the pointed electrode is ideal for high-quality and precision welding. TIG welding is especially useful for welding thin materials, but requires a high level of skill and can only be performed at low speeds. Similar to MIG welding, TIG welding requires little to no finishing work.

Plasma arc welding is a process that is similar to TIG welding. However, plasma welding has a number of advantages over TIG welding which makes it an interesting alternative to laser welding, especially on sheet and other

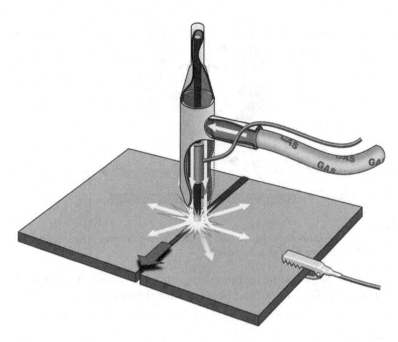

FIGURE 6.1
MIG welding fundamentals [4].

components with a sheet thickness of up to 8 mm. The difference between plasma arc welding and TIG welding is that the arc consists of plasma (a gas with positive ions and electrons) (Figure 6.3). The plasma arc is constricted with the help of a water-cooled, fine-bore copper nozzle which squeezes the

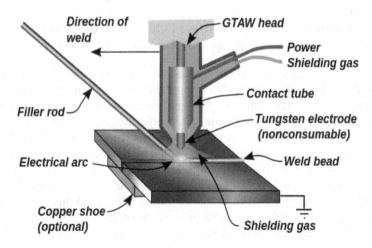

FIGURE 6.2
TIG welding fundamentals [5].

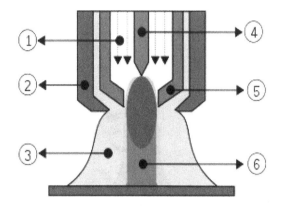

FIGURE 6.3
Plasma welding fundamentals [6]. 1. Gas plasma, 2. Nozzle protection, 3. Shield Gas, 4. Electrode, 5. Nozzle constriction, 6. Electric arc

arc, increases its pressure, temperature, and heat intensity. These factors improve arc stability, shape, and heat transfer characteristics.

6.1.2 Beam Welding

There are two types of beam welding processes: laser beam welding (LBW) and electron beam welding (EBW). LBW is a fusion welding technology using a laser beam as the primary heat source. The laser beam provides a localized high-power density (of the order of $1 \; MW\!\!\big/\!\!_{cm^2}$ or more) allowing for a narrow deep weld and high welding speeds. The localized heat input in laser welding leads to small heat-affected zones and results in high heating and cooling rates.

EBW is a joining method which uses a tightly focused beam of high energy electrons that strike the workpiece with a power density of $10^5 \; W\!\!\big/\!\!_{mm^2}$ or greater. The high-power density causes vaporization of the molten metal, leading to the formation of the "welding capillary" or "keyhole" that is characteristic of EBW. EBW results in extremely narrow, deep-penetration welds with a minimal heat-affected zone, thus requiring a minimal power input. Additionally, the bulk of the assembly remains cold and stable. In the EBW process, the beam is produced and controlled by the electron beam generator. The electrons emerge from the cathode consisting of a tungsten filament heated to approximately 2500 °C. Voltages of up to 150 kV between the cathode and anode accelerate the electrons toward the workpiece. An electromagnetic lens focuses the diverging electron beam to a spot with high-power density. When the electrons hit the surface of the workpiece, their kinetic energy is mostly transformed into heat. Only a small part of the energy is emitted as X-rays.

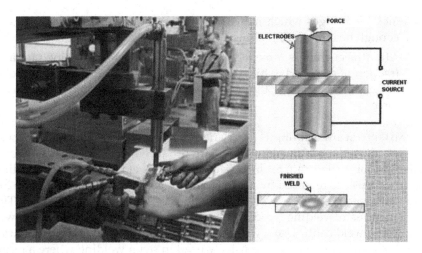

FIGURE 6.4
Spot welding [7].

However, compared to the widespread applications of LBW, EBW has only found limited areas of application. The main advantages of LBW are that LBW can be transmitted through the air (no need for a vacuum), LBW can be easily automated, there are lower requirements for occupational safety, and the laser beam can be transmitted through fiber-optic cables in order to share or switch between workstations.

6.1.3 Electric Resistance Welding

Electric resistance welding refers to a group of thermoelectric welding processes, such as spot and seam welding. The weld is made by conducting a strong current through the metal to heat up and, finally, melt the metal at a localized point (Figure 6.4). This point is predetermined by the design of the electrodes and/or the workpiece being welded. A force is always applied before, during, and after the application of current to confine the contact area at the weld interfaces and, in some applications, forge the workpieces.

Resistance spot welding is a process that utilizes the heat obtained by the resistance to the flow of electric current to the workpieces through electrodes that concentrate current and pressure to the weld area. The generated heat is used to melt and solidify a nugget at the faying surfaces of a joint.

The three main stages of spot welding are as follows (Figure 6.4):

1. The electrodes are brought in contact with the surface of the metal and slight pressure is applied.
2. A high current flows for a short time. As it passes through the bulk metal, the weld current is distributed over a large area. However, as it approaches the interface, the current is forced to flow through the

metallic bridges which increases the current density and generates enough heat to cause melting.

3. After the current is removed, the electrode force is maintained for a fraction of a second, while the weld rapidly cools. At the end of the cycle, solidification of the nugget is completed under the electrode force.

Resistance seam welding is a variation of the spot welding process where a series of overlapping weld nuggets are produced that form a continuous, leak-tight joint (Figure 6.5). In resistance seam welding, the spot welding electrode tips are replaced by a pair of driven copper wheels (typically ~200 mm in diameter) or one wheel acting against a stationary backing piece. The resistance seam welding process depends on three parameters: the power supply and weld control unit, weld wheel configuration, and sheet configuration. A consumable wire is extensively used in seam welding where a properly shaped copper wire is fed between the wheels and materials to be joined in order to provide consistent clean contact. Robots are extensively used in precision welding processes. Material on this topic is widely available in the literature such as in reference [8].

6.1.4 Brazing

Brazing is a joining method which provides a permanent bond between the parts to be joined with the help of a brazing filler metal. The composition of the filler alloy is such that its melting point is slightly below the melting

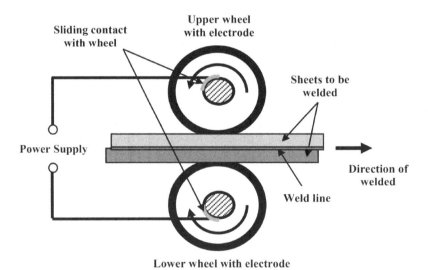

FIGURE 6.5
Resistance seam welding (adapted from ref [9]).

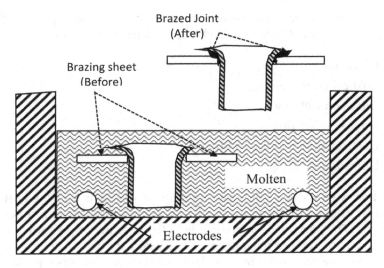

FIGURE 6.6
Flux-tip brazing (adapted from ref [10]).

range of the parent metal of the parts. Brazing is distinguished from welding by the fact that the parent metal does not melt during the process. Brazing differs from soldering because the brazing filler metal is an aluminum base alloy and the working temperatures for soldering are appreciably lower.

Modern techniques of brazing aluminum have been established as an important mass production method over the past decades. The three main types of brazing include torch brazing, flux-dip brazing (see Figure 6.6), furnace brazing, and vacuum/controlled atmosphere brazing.

Some advantages of brazing include:

- Joining components with very small thicknesses
- Joining aluminum alloys to dissimilar metals
- Lower temperatures than welding
- Low component distortion
- Accommodates large-scale joint areas
- Meniscus surface formed by the filler metal is ideally shaped for good fatigue properties
- Low finishing costs

6.1.5 Solid-State Welding

Solid-state welding describes a group of joining technologies which produce coalescence at temperatures below the melting point of the parent materials without the addition of a third material. External pressure and relative movement may or may not be used to enhance the joining process. Some common examples of solid-state welding include:

- Friction (stir) welding
- Cold pressure welding
- Diffusion welding
- Explosion welding
- Electromagnetic pulse welding
- Ultrasonic welding

Friction welding is a solid-state welding process that uses the heat generated by friction (induced by relative mechanical movement) to join two materials. The parts to be joined are held together under pressure. Friction welding techniques are generally melt-free where the base materials are kept under their melting temperatures. The frictional heat causes a plastic zone (soft interface) between the parts to be joined. The applied external force presses the parts together and thus creates the joint. The combination of short processing times and the development of the heat directly at the interface results in fairly narrow heat-affected zones, also caused by upsetting a portion of the interface out of the weld joint during the process.

Rotational friction welding (Figure 6.7) is a frictional welding process that involves rotating one part against a stationary component to generate the friction heat at the interface. When a sufficiently high temperature has been reached, the rotational motion ceases and additional pressure is applied to join the materials together.

Linear friction welding is similar to rotational friction welding except that the moving part oscillates laterally relative to the other part instead of rotating (Figure 6.8). Linear friction welding is a high-quality joining process that creates a solid-phase bonding with the parent material properties.

Diffusion welding is a solid-state welding process by which two metals (usually dissimilar) can be bonded together. The necessary diffusion processes involve the migration of atoms across the interface due to the existing concentration gradients. The two materials whose surfaces must be machined as smooth as possible and kept free from contaminants are pressed together (at elevated temperatures usually between 50% and 70% of the melting point). The pressure is used to relieve the void that may occur due to the different surface topographies. The process does not involve plastic deformation, melting or relative motion of the parts. Filler metal may or may not be used. Once clamped, pressure and heat are applied to the components, usually for many hours; preferably under vacuum or inert atmosphere. When a layer of filler material is placed between the faying surfaces of the parts being joined, the term "diffusion" is generally used.

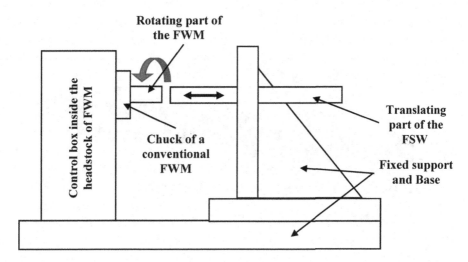

FIGURE 6.7
Schematic of a rotary friction welding machine for joining round parts (FWM) (adapted from ref [11]).

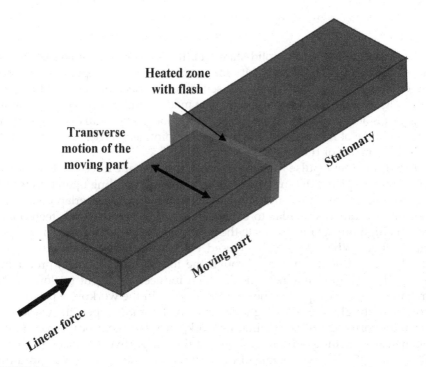

FIGURE 6.8
Linear friction welding (adapted from ref [12]).

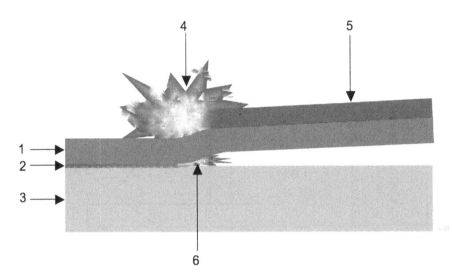

FIGURE 6.9
Explosion welding. [1. flyer (cladding). 2. resolidified zone (needs to be minimized for welding of dissimilar materials). 3. target (substrate). 4. explosion. 5. explosive powder. 6. plasma jet [13]].

Explosion welding is a solid-state welding process where coalescence is accomplished by the high-velocity impact of one of the components onto the other part. Figure 6.9 [13] shows the schematic of the explosion welding process. Figure 6.10 [14] shows a sample part made using this process. The moving part is accelerated by a controlled detonation of chemical explosives. Due to the nature of this process, the producible joint geometries must be simple (typically plates or tubes).

Electromagnetic pulse forming uses electromagnetic forces to deform and join the workpieces (Figure 6.11). It is an automatic welding operation which can be used for tubular and sheet metals placed in the overlap configuration. It is a process similar to explosion welding because both techniques rely on high impact rate to create the bond and the joint boundary displays a ripple effect. When the switch is closed, electrical energy stored in the capacitor bank (left) is discharged through the forming coil (orange) producing a rapidly changing magnetic field which induces a current to flow in the metallic workpiece (pink). The current flowing in the workpiece produces a correspondingly opposite magnetic field which rapidly repels the workpiece from the forming coil, reshaping the workpiece. The reciprocal forces acting against the forming coil are resisted by the supportive coil casing (green). Reference [17] discusses some of the topics on friction, explosive and ultrasonic welding processes.

Ultrasonic welding creates a solid-state weld by the local application of high-frequency vibrations as the workpieces are held together under pressure

FIGURE 6.10
Polished section of an explosion weld with typical wave-structure [14] (By Kat1100 – own work, CC BY 3.0).

(Figure 6.12). This is a cold welding process since the heat generated by the ultrasonic energy is not essential to the formation of the joint. The welding occurs when the ultrasonic tip that is clamped against the workpieces oscillates in the plane parallel to the weld interface.

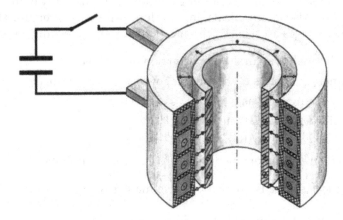

FIGURE 6.11
Compression of a tubular workpiece by electromagnetic forming [15].

FIGURE 6.12
Ultrasonic welding of thin foils at the Essen Welding Fair. The sonotrode is rotated along the weld seam [16].

6.2 Mechanical Joining

This section intends to show different aspects of joining aluminum using mechanical joining which is one of the most widely used joining techniques especially for aluminum components. Example: In order to reduce weight on the car body, the usage of lightweight materials in the body-in-white application has increased a lot. Thus, the different type of mechanical joining techniques becomes important.

All mechanical joining methods are cold forming techniques. One general benefit for all mechanical joining methods is the mobility of the material after joining. Also, mechanical joining offers impressive strength. There are certain advantages associated with mechanical joining techniques, such as:

- No thermal structural transformation of workpieces
- Numerous choices of the material of rivets, sizes, form, etc.
- High-strength capacity
- Mobility of material after joining
- Easy-to-control quality
- Good environmental behavior – no emission or pollution

Mechanical joining is divided into two main groups:

- Mechanical joining without additional fastener
- Mechanical joining with additional fasteners

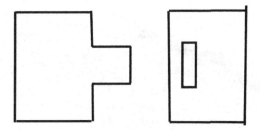

FIGURE 6.13
Basic illustration of mechanical joining without additional fasteners. (Raghu redrew this; adapted from ref [18–20]).

6.2.1 Mechanical Joining Without Additional Fasteners

In this type of mechanical joining, the joint is produced at the point of contact directly out of the sheet material components – meaning without any additional fastener (Figure 6.13). One or more tabs (also called legs, ears, or tongues) are formed on one of the workpieces as shown in Figure 6.13. Matching slots slightly longer than the tab widths are punched in the mating part. Both features are made during the press operations that fabricate the parts.

This type of technique can be used in two ways:

- Hemming
- Clinching

6.2.1.1 Hemming

Hemming is a forming operation in which the edges of the sheet are folded or folded over another part in order to achieve a tight fit (Figure 6.14). Normally, hemming operations are used to connect parts together, to improve the appearance of a part and to reinforce part edges. In car parts production, hemming is used in assembly as a secondary operation after deep drawing, trimming, and flanging operations to join two sheet metal parts (outer and inner) together. Typical parts for this type of assembly are hoods, doors, trunk lids, and fenders.

The accuracy of the hemming operation is very important since it affects the appearance of the surface and surface quality. Material deformations, which occur during the hemming process, can lead to dimensional variations and other defects in parts. Typical hemming defects are splits or wrinkles in the flange, material overlaps in the corner areas or material roll-in. Therefore, it is important to use simulation tools in order to, on the one hand, better understand the hemming process and, on the other hand, significantly reduce the number of "trial and error" loops during try out and production.

FIGURE 6.14
Simulation of conventional die hemming (courtesy of AutoForm Engineering [21]).

There are two types of hemming operations:

- Conventional die hemming
- Roll hemming
- Conventional die hemming

Conventional die hemming is suitable for mass production. In this process, the flange is folded over the entire length with a hemming tool. Normally, the actual hemming is a result of a forming operation in which the flange is formed with a hemming tool after the drawing and trimming operations have been completed. The formed flange is then hemmed in several process steps. These steps include, for example, the prehemming and final hemming depending on the respective opening angle of the flange. Production plants for conventional die hemming are very expensive, but the cycle times are very low. Figure 6.14 shows the virtual simulation of hemming operation using Autoform software.

- Roll hemming

Similar to die hemming, roll hemming (Figure 6.15) is carried out incrementally with a hemming roller. An industrial robot guides the hemming roller and forms the flange. Roll hemming operation can also be divided into several prehemming and final hemming process steps. Roll hemming is very flexible to use and tool costs are significantly lower as compared to those of conventional die hemming. However, the cycle times are much higher since the hemming is realized using a hemming roller which follows a defined path.

6.2.1.2 Clinching

In technical terms, clinching is defined as a single or multistep fabricating process with a common displacement of the materials to be joined combined

FIGURE 6.15
Simulation of roll hemming (courtesy of AutoForm Engineering [21]).

with local incision or plastic deformation and followed by cold compression, so that a quasi form locking joint is produced by flattening or flow pressing [22]. Clinching opens new possibilities of joining lightweight sheet materials in the assembly field of lightweight structure manufacturing. The uninterrupted action of cold-forming produces the joint element at the clinched point directly out of the sheet material components. Lightweight sheet materials normally include low-density metals, such as aluminum, and nonmetals, such as plastics and various composites. Metal–metal combinations can be joined by conventional clinching, but some metal–nonmetal pairs can only be connected by hybrid clinching or modified clinching. Figure 6.16 shows a typical clinching machine.

Depending on the tools used, clinching can be classified into:

- Round clinching
- Square clinching

Sheet materials are only deformed by round clinching, as shown in Figure 6.17. There are different phases of this process. Different clinching dies are shown in Figure 6.18, which includes rectangular or square clinching. Both deformation and cutting of sheet are required in square or rectangular clinching. Thus, the fatigue properties will be affected, and the water tightness does not apply to square clinching. Both round clinching and square clinching are not recommended for brittle materials as mechanical clinching is generally a cold-forming process, whereas it can be easily used for aluminum joining as it can withstand very low temperatures.

FIGURE 6.16
Clinching process [23].

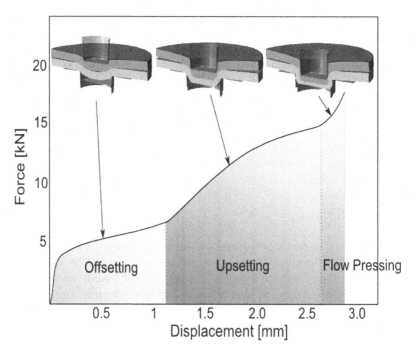

FIGURE 6.17
Clinching process [24].

FIGURE 6.18
Clinching dies [25].

Compared to the more traditional sheet materials joining methods, such as spot welding, spot friction welding, and self-piercing riveting, the advantages of clinching include:

- The ability to join multiple similar or dissimilar sheet materials
- Improved fatigue properties of the product
- Ease of process automation and monitoring
- Environmental safety

However, as with any sheet materials joining technology, clinching has some disadvantages:

- Relatively high force required for the forming process
- Inapplicability to brittle sheet materials

6.2.2 Mechanical Joining With Additional Fasteners

Mechanical fastening continues to be the dominant method for joining large composite parts, despite the obvious disadvantages of drilling holes in composites. A thorough understanding of bolted joint behavior is thus essential to the design of efficient structures from composite materials. Mechanical fastening includes joining components and subassemblies by press-fitting, snap fitting, and using fasteners, such as threaded fasteners and rivets. Threaded fasteners are separate parts that have internal or external threads and include bolts, nuts, and screws. Rivets, which consist of a head and body, are used to fasten two or more parts together by passing the body through a hole in each part and then forming a second head on the body end. Some advantages of mechanical joining include:

- Virtually any material in any shape can be joined by mechanical fastening, given enough ingenuity
- Especially good for joining different materials (e.g., composite to metal)
- Joint quality is reliable and readily determined, given enough operator skill. However, mechanical joining usually reduces fatigue life
- Essential where two parts will move relative to each other (e.g., hinges for doors)
- The nonpermanence of many fasteners is useful for products that may need repair/maintenance or need access to the interior

6.2.2.1 Screws and Bolts

Screw joints are detachable joints which require either a separately manufactured mating thread or an extra, internally threaded component (nut). A distinction can be made between connections formed using a clearance hole (bolts) and internally threaded holes (screws). Since aluminum alloys show relatively low compressive strength, the contact surfaces must be generally protected by washers under the screw and the nut.

Threaded fasteners are widely used and manufactured in a wide variety of shapes and sizes. Application methods range from manual use to fully automated (robotic) systems. For aluminum structures, screws and bolts are usually made of steel, but also other materials (including high-strength aluminum alloys) can be used.

- Bolted connections, such as the one shown in Figure 6.19, may be produced by simply bolting through the aluminum parts. If bolting through a closed section, it may be necessary to provide internal support to prevent the section from collapse under high installation loads.

A special benefit of the aluminum extrusion technology is the possibility to integrate continuous tracks for nuts or bolt heads into the cross section of the profile enabling stepless fastening without any need to machine the profile.

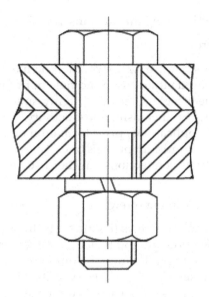

FIGURE 6.19
Vertical cut-away view of a tension-loaded bolted joint [26].

Using special nuts/bolts, fastening can even take place without having to slide in the nut/bolt from the end of the track.

- Female threads in aluminum components

The load-bearing resistance of a joint formed between a bolt and a machined thread depends not only on the strength of the material, but also on the form of the thread and the area of the mating surfaces. When the threads are cut or cleanly grooved into wrought aluminum components, the screws can be undone and tightened up repeatedly without damaging the thread. In low-strength aluminum alloys and aluminum castings, steel thread inserts may be used to increase the pullout force and facilitate assembly and disassembly.

- Threaded studs and nuts

Bolted connections can also be made with threaded studs and nuts which are previously fixed to an aluminum component [27]. Such solutions are applied when tapped threads are not possible due to small wall thicknesses or when the material is too soft to support tapped threads. The applied threaded elements take on the role of either the nut or the bolt and enable the attachment of further components in a second step. The selection of the insert type depends upon the required strength and whether access is possible from one or both sides. Aluminum threads may be applied for lightly loaded connections, whereas steel-threaded studs and nuts are preferred for applications where higher strength or frequent disassembly is necessary.

An example of a self-tapping threaded insert is shown in Figure 6.20.

* Self-tapping screws

Self-tapping screws (Figure 6.21) form their own threads when screwed into core holes prepared either as a blind hole in full material or as a pre-punched hole in sheet metal. They are highly suited for joining thicker aluminum components, increasing productivity, and reducing joining cost. For thin sheet applications, prior formation of a rim should be considered. In extruded aluminum profiles or castings, screw ports for transverse and longitudinal connections can be directly integrated into the component.

* Hole and thread-forming screws

Hole and thread-forming screws (Figure 6.22) eliminate the drilling operation and enable high-strength joints due to the increased thread engagement in the formed draught. Joining of aluminum components with up to 5 mm thickness is generally possible. For specific material combinations, a pilot hole may be advantageous. The usual tolerance problems as overlapping of draught and insertion hole do not apply. One-sided accessibility provides also for an assembly into hollow profiles without any counter-support. Hole and thread-forming screws are highly suited for automated assembly. Stainless steel screws are most often used.

FIGURE 6.20
Self-tapping threaded insert [28] (Wikimedia).

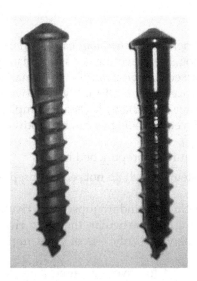

FIGURE 6.21
Self-tapping screws [29].

There are essentially two different methods:

- Cold hole and thread-forming screws exhibit a sophisticated screw design. The special geometry of the screw point produces a high-contact pressure per unit area which then leads to the necessary plastic deformation of the material.
- In the flow-forming (drilling) process, a tapered but unthreaded punch rotating at high speed is forced down to pierce through the metal. The sheet metal heats up and a collared hole is formed by plastic deformation. A thread can then be tapped into the cylindrical hole.

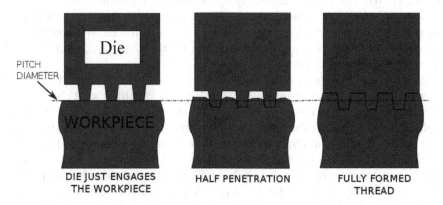

FIGURE 6.22
Simple concept of thread forming by rolling [30].

6.2.2.2 Riveting

Rivets are somewhat permanent mechanical fasteners which clamp two or more material layers together. Riveting is a safe and easy-to-apply technique, also applicable to mixed material joints. Pneumatic, hydraulic, manual, or electromagnetic processes are all highly effective in driving the rivets. Assembly systems range from hand tools and simple workstations to fully automated systems. Rivet technologies can be subdivided into two groups:

- Rivet systems requiring prepunched holes
- Self-piercing systems which do not require prepunched holes

The first category includes standard (upsetting) riveting systems (solid riveting and blind riveting). In particular, the blind riveting process – which can be applied from one side only – is of great importance in automotive applications.

Assembly systems used for riveting range from hand tools and simple workstations to fully automated systems. Pneumatic, hydraulic, manual, or electromagnetic processes are all highly effective in driving the rivets.

- Solid Rivets

The application of solid rivets requires prepunched or predrilled holes as we all as two-sided access. Before being installed, a solid rivet consists of a cylindrical shaft (or shank) with a head on one end. Once the rivet has been inserted, the closing head is formed from the rivet shank by plastic deformation. Because there is effectively a head on each end of an installed rivet, it can support tension loads; however, it is much more capable of supporting shear loads. With all-aluminum constructions, cold-formed aluminum rivets are used almost exclusively.

Rivets are classified according to the shape of the rivet head and the form of the shank. The most common types of solid rivets show a round or flat, sometimes also countersunk, heads. Apart from solid shanks, semitubular or tubular shanks are also used in order to reduce the closing forces. Joint characteristics can vary greatly depending on the rivet type, material, and geometry. Figure 6.23 shows a solid rivet.

As shown in Figure 6.24, the rivet forming process, in this case, the orbital riveting process, involves multiple stages and the resulting joint characteristics depend on the type of the process and the rivet shank. The shank on a solid rivet expands in the hole during the riveting process, typically forming an interference fit. On a semitubular rivet, where the part of the shank which protrudes beyond the back of the second workpiece is hollowed out, the hollow tenon curls over on impact, drawing the parts together with minimal shank swell. Semitubular or tubular rivets are thus ideal to use as pivot points since the rivet only swells at the tail.

FIGURE 6.23
Solid rivets [31].

With all-aluminum constructions, cold-formed aluminum rivets are used almost exclusively. Hot-formed steel rivets are only used for joining aluminum and steel. However, care must be taken to avoid the negative effects of the rivet heat on the properties of the aluminum component. Solid rivets are

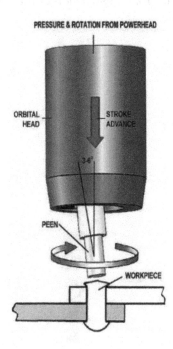

PRESSURE & ROTATION FROM POWERHEAD

ORBITAL HEAD

STROKE ADVANCE

3-6°

PEEN

WORKPIECE

FIGURE 6.24
Example of a 4-step orbital rivet forming process [32].

pressed through the two materials and into a solid die. When they hit the die, the penetrating end deforms and spreads out. This creates a permanent hold since the head and the deformed tail of the rivet are both larger than the hole in the material. Once in place, the only way to remove a rivet is to cut it from the workpiece.

- Blind Rivets

As shown in Figure 6.25, a blind rivet consists of two components, a smooth, cylindrical rivet body and a solid rod mandrel with a head which runs through the hollow rivet shaft. For installation, the rivet is placed into an installation tool and inserted into the prepunched hole. The tool pulls the mandrel into the rivet body and the material layers, the rivet walls are expanded and firmly compressed in the hole while a tightly clinched load-bearing area is formed on the reverse side. The upset head on the rivet body securely clamps the material layers together. Finally, the mandrel reaches its predetermined breakload, the spent portion of the mandrel breaks away and is removed. The remaining portion of the mandrel is captured inside the sleeve and plugs the opening in the rivet shell. The entire installation cycle takes about one second. Blind rivets are available in different designs both for nonstructural and structural applications. Most important is the controlled expansion of the breakstem rivet body. This is achieved through

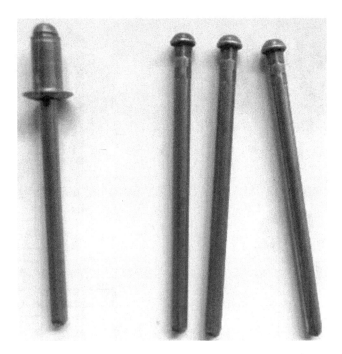

FIGURE 6.25
Examples of blind rivets [33].

an appropriate mandrel design and selection of the rivet material. In aluminum structures, aluminum rivets are normally used. Steel mandrels can be chosen for strength reasons; stainless steel is the preferred option, but there are also steel mandrels with protective coatings.

- Lockbolts

Lockbolts (Figure 6.26) require pre-existing holes and two-sided access. They allow the realization of high-strength joints with a high, controlled clamp, which will not work loose, even during extreme vibration. Lockbolts consist of a pin which is inserted in the hole and a collar which is placed on the pin from the opposite end. The tool is placed over the pintail and when activated, the pinhead pulls against the material, the tool anvil pushes the collar against the joint, and the initial clamp is generated. The tool then swages the collar into the pin, the pintail breaks, and the installation is complete.

The shear strength of lockbolts varies according to the material strength and minimal diameter of the fastener. By increasing the diameter or selecting a higher strength material, the shear strength of the fastener can be increased. The tensile strength of lockbolts is dependent on the shear resistance of the collar material and the number of grooves it fills. As with rivets, there are multiple steps involved in the installation of lockbolts. In the first step, the pin is placed into the prepared hole and the collar is placed over the pin. In the initial stage of the installation process, the tool engages and pulls on the pintail. The joint is pulled together. At the same time, the cone-shaped

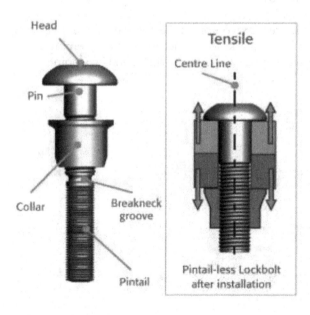
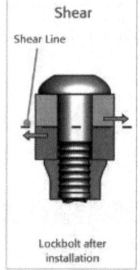

FIGURE 6.26
Examples of Huck LockBolts (courtesy of Huck International) [34].

anvil is forced down the collar, pushing the collar against the joint and generating the initial clamp. In the second step, the tool swages the collar into the grooves of the harder pin. The squeezing action reduces the diameter of the collar, increasing its length. This, in turn, stretches the pin, increasing the clamp force over the joint. When the collar is fully swaged, the pin breaks and the installation is complete. Lockbolts manufactured by Huck International offer many advantages including using less number of bolts to achieve lightweighting, easy installation, and providing tight joints that protect from various load and environmental conditions [34].

The installation steps of a Lockbolt are shown in Figure 6.27.

Step 1

Pin placed into prepared hole; Collar placed over pin

Step 2

Tool is placed over the fastener pintail and activated; Pinhead pulled against material; Anvil pushes collar against joint; Initial clamp generated

Step 3

Tool swages collar, increasing clamp

Step 4

Pintail breaks, installation complete

Self-piercing Semi-Tubular Rivets

Self-piercing riveting (Figure 6.28) combines the hole-cutting and riveting process. Two-sided access to the workpiece is necessary. Self-piercing rivets eliminate the need for alignment and minimize distortion. The joints need

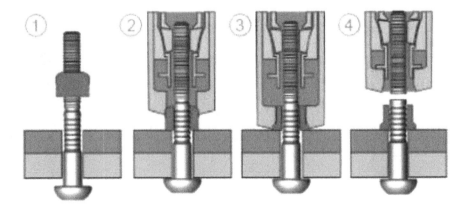

FIGURE 6.27
Installation steps of a lockbolt (reproduced with permission of Huck International) [34].

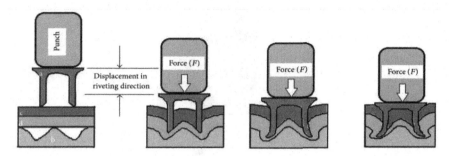

FIGURE 6.28
Examples of hollow self-piercing rivets [35].

to be of a lap-type configuration. In tensile and peel loading, self-piercing rivet joints have virtually the same static strength as spot-welded joints but show higher strength and stability under dynamic load. In terms of part size or configuration, the only condition is that the rivet actuation cylinder and C-frame can access the joint. Assembly equipment can be stationary, robotic or integrated into an assembly cell. Semitubular rivets pierce the top sheet, whereas the lower material layer is not penetrated. A shaped die on the underside reacts to the setting force and causes the rivet tail to flare within the bottom sheet. This produces a mechanical interlock and creates a button in the bottom sheet. For the best results, the rivet is applied from the direction of the thin into the thick sheet, or from the low-strength into the high-strength material. If this is not possible, it is recommended that the bottom layer thickness is not less than one-third the joint stack thickness. Semitubular rivets can fasten stacks of two or more aluminum layers up to 12 mm total thickness. The joint is leak-proof and has a very high degree of integrity.

Although aluminum self-piercing rivets are available, steel elements covered with a protective layer to prevent galvanic corrosion are generally used.

• Self-piercing Solid Rivets

Self-piercing solid rivets pierce through the whole material stack and the punched-out parts must be removed (similar to Figure 6.28). Several research papers are available on self-piercing solid rivet technology for joining dissimilar metals. Two process variants are used. For rivets without countersunk heads, the rivet is locked by the surrounding material under the compressive action of the shoulders both on the punch and the die. Rivets with a countersunk head show one or more grooves in the rivet shaft and a ring-shaped contour on the die plate presses into the bottom material layer to create the undercut necessary for the connection strength while the punch side remains flat. Nevertheless, the resulting joint strength is inferior to that of semitubular riveted joints. Two or more material layers can be joined up to a combined sheet thickness of approximately 9 mm. Aluminum solid punch

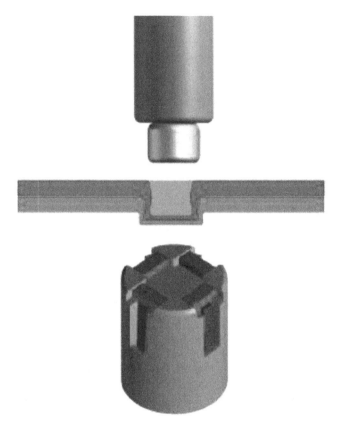

FIGURE 6.29
Example of clinch riveting [36].

rivets can be even reworked mechanically. Different form-locking methods when self-piercing riveting with solid rivets are discussed in the next section.

- Clinch Riveting

The clinch riveting technology shown in Figure 6.29 complements the clinching process (Figure 6.17) with an additional retaining member. A simple cylindrical rivet is pressed-in and formed during the clinching process. Just like with the clinched joint, the materials to be joined are not cut, but only deformed inside a die cavity. The result is a joint highly suited for shear loading even when used with thin materials.

- Tack High-Speed Joining

High-speed bolt setting is a simple and fast joining process which requires no prepunched holes and only one-sided access. But there is a need for a relatively stiff counterpart, i.e., the preferred application is a sheet/profile joint. The stack-up range can be from 3 mm up to 6 mm and the setting time is a second or less. That is why, it is a high-speed joining process. Figure 6.30 shows an example of a RIVTAC® rivet [37], which can be ideally used for complex geometries, such as profiles, and complex cast materials. This process is also optimal for combining with adhesive. A nail-like fastener ("tack") is accelerated to high speed and driven into the parts to be joined (Figure 6.31). There are four steps in this joining process: (1) Clamping, (2) Entering, (3) Penetration, and (4) Bracing.

The speed, which can be controlled via the adjustable pressure, is optimized to suit the material type and wall thickness. The pointed tip of the tack displaces the material without forming a slug but leads to a momentary local temperature rise. The material flowing in the joining direction forms the draught, whereas material flowing contrary to the joining direction flows into the knurled shaft of the tack and leads to a high form fit. Joint stability in the lower joint section is achieved by a combination force fit, resulting from the restoring force of the displaced material, and form fit.

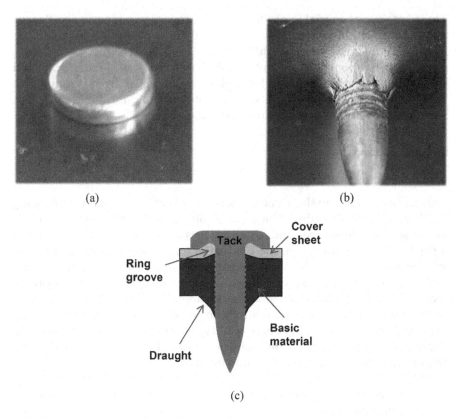

(a)

(b)

(c)

FIGURE 6.30
Example of RIVTAC joining (a) tack rivet head, (b) tack tail, and (c) different parts of the joint [37] (courtesy of Boellhoff, Inc.).

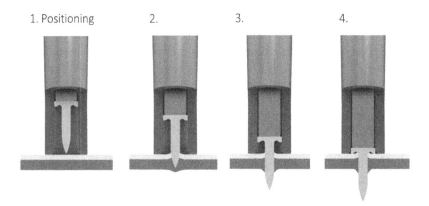

FIGURE 6.31
Example of RIVTAC high-speed joining process [37] (courtesy of Boellhoff, Inc.).

6.2.3 Special Mechanical Joints

There are several special joints, for example, to join two pipes using flanges and clamps, and the different ways to join them to minimize or avoid leakage from the joint. Other special joints are used for steering mechanisms, suspension components, transmission shafts, etc. Most of these use ball and socket joints or universal joints. Examples of front-wheel drive assembly and ball joints used for suspensions components are shown in Figure 6.32 [38]. Details of these types of joints are not discussed here.

6.3 Adhesive Bonding

This section intends to show one or more aspects of joining aluminum using heatless, adhesive bonding. The view of joining techniques can be used to assess specific advantages and disadvantages of adhesive bonding which is a further key technology for the binding of joined aluminum components. Joining is a key issue in the development and manufacturing of products and building structures. To make optimum choices among the numerous joining methods, engineers have to define the requirements of the joint with respect to the material combinations (aluminum, mixed metals, organic, coated/ uncoated), design (thick, thin, accessibility), functionality (strength, durability, removability), as well as the characteristics of the joining method. Some important factors to adhesive bonding include:

- No structural or geometrical change due to thermal impact ("cold" connection process)
- Wide range of possible connections (metal to plastic, organics, coatings, etc.)

FIGURE 6.32
Typical front-wheel drive system [38] (courtesy of Mark Brown from Hampton, New Brunswick, Canada).

- Possible combination of adhesive bonding with mechanical or thermal joining techniques
- High exploitation of material properties because of surface-related force transfer

6.3.1 Design Aspects

Figure 6.33 shows the different layers of a typical aluminum sheet. The top few layers except the base material are very thin of the order of microns. The design of an adhesive bond may be simple or complicated depending on the adhesive's function. For optimum performance of the materials and adhesive, some general principles should be considered:

- Stress in the direction of maximum strength
- Maximize bonding surface
- Adhesive applied uniformly
- Adhesive is thin and continuous
- Minimize stressed areas

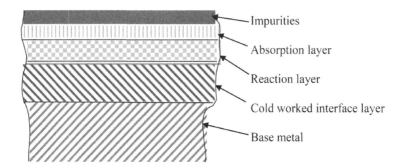

FIGURE 6.33
Typical aluminum sheet surface layers (redrawn and adapted from ref [39]).

It is important to know the duration, direction, and load of the forces being applied to a designed joint. Most adhesives used for structural purposes are relatively strong in shear. Conversely, these same adhesives have relatively low adhesive strength in tensile or peel. The design of the bond line should take into consideration the forces that occur when the device is under load. A simple lap shear bond has peel forces at the ends while the center observes little stress.

When designing adhesive-bonding applications, optimizing joint design is an important consideration. Adhesive joints are not geometrically limited the way their mechanical fastener counterparts are. This leaves designers free to focus on the various mechanical and chemical stresses. A specific joint is expected to withstand at its anticipated service temperature range. During the design phase, particular attention must be paid to the potential effects of mechanical shock and vibration, especially in dynamic bonding applications. Furthermore, assembly, manufacturing methodology, and cost factors must all be considered when proposing a joint design.

An understanding of the various possible joint designs for application is an essential step to finding the optimum bonding solution. Typical joint design and stress development data offer designers several choices. The following shows the most widely employed types of joint designs in use today (Figure 6.34):

The butt joint is the simplest design. It is simply bonding two parts end to end. Generally, butt joints are not recommended for applications where

FIGURE 6.34
Types of adhesive joint design (Raghu redrew this; adapted from ref [39]).

available surface area is less than ½ to 1 inch, such as thin films, sheets and fibers. Bonding strengths for properly designed butt joints can vary from 800 psi in excess of 3,000 psi.

Scarf joints are similar but have the joints leveled at matching angles to enhance the surface area available as well as increase the shear resistance. Shear stresses set in a plane with two substrates moving in opposite directions occur very often. Most structural adhesives can withstand 2,000–3,000 psi of shear stress at ambient temperatures.

In a lap joint, two substrates are joined by bonding together large surfaces of each piece. Bonding the same substrates by using the butt or scarf joints would result in less surface area being joined than a lap joint. However, while lap joints allow more bonding area, they can result in offset surfaces being susceptible to peel, which occurs when one of the substrates is deformed and pulled away from the bond line of the other substrate. Less cleavage stresses develop when tensile forces are unevenly applied to one edge of a joint, forcing it to open. Both cleavage and peel failures decrease the effective surface area of the bond and can initiate "unzipping" between the substrates as pressure continues to be applied to the joint.

For lap joints loaded in tension, the load is transferred predominantly by shear stresses in the adhesive layer (Equation 6.1). If the adhesive was loaded uniformly then

$$\tau_a = [P / (L * b)] \tag{6.1}$$

Where:

τ_a is the adhesive shear stress

P is the load

L is the overlap length

b is the width

6.3.2 Adhesive Selection

Selecting the proper adhesive involves consideration of:

- Manufacturing conditions
- Substrates to be bonded
- End-use environment
- Cost factors

The above-mentioned selection factors are explained in detail as follows:

- Manufacturing conditions include machinery, materials-handling methods, and plant conditions.
- Machinery: The design engineer should determine what machinery is already in place at the manufacturing facility. The engineer

may recommend the appropriate applicators for the type of adhesive required. Spray and extrusion applicators are excellent for applying low-viscosity fluid adhesive products. Roller applicators normally apply medium-viscosity adhesives. Pot-applicators are suited to high viscosity.

- Material-bonding methods: Machinery is not limited to applicators. The manufacturer's materials handling is also important. For example, how much time does the manufacturing process allow for bonding? Does product handling require the initial bond strength to be greater than normally needed? Is there time for a product which bonds more slowly?

- Plant conditions: Manufacturing conditions involve more than the machinery and materials-handling methods in the plant. Plant conditions, such as the condition of the equipment and the skill of the product personnel should also be considered.

- Substrates to be bonded: The openness, or porosity, of the substrate to be bonded can place additional demands on an adhesive system. Excessive penetration, hardness, or impenetrability can make some adhesives unsuitable.

Adhesion to a coated surface, such as painted or plated steel, must consider the surface coating, not only the base substrate. The coating has a profound effect on whether certain adhesive systems will be suitable for use in that application.

- For example, many molded plastics have residual mold-release agents on their surface, and most attempts to bond these plastics will fail unless the surface is cleaned. A solvent wiping usually is adequate to render the surface bondable with the most appropriate adhesives.

Likewise, some base metals quickly oxidize, so the surface they provide is not really metal, but a metal oxide. However, many surfaces can be treated to achieve more suitable levels of adhesion.

It is essential to know what the coating is. Substrate characteristics place requirements on the selection of the proper handling system.

- End-use environment: The end-use environment includes all conditions to which the adhesive bond will be subjected during the useful life of the product. Considerations vary with the application and include stress, the kind of joint being used, temperatures, exposure to moisture, flexibility, age, stability, and aesthetics.

- Stresses: These include stress placed on the glue line during construction as well as end-use stress – the conditions the construction will be expected to encounter.

There are four types of stress:

- Shear, when the adherends move in parallel planes
- Tension, when the adherends are pulled apart in the same plane
- Cleavage, when two adherends are pried apart at the end of a lap joint
- Peel, an exaggerated form of cleavage that occurs when a flexible adherend is bent away from the bond line

There are three bonding failure types:

- Adhesive: The adhesive peels off the substrate surface.
- Cohesive: The adhesive sticks to the substrate surface but rips itself apart.
- Delamination: The adhesive sticks together, and to the substrate surface, but pulls the coating off the metal.
- Joints: There are many kinds of joints including butt, lap, beveled lap, scarf lap, and invert-T. Generally, adhesive-bonded joints in load-bearing structures should be loaded essentially in shear, minimizing the stress induced by peel, cleavage, and impact forces.

Joints must be designed specifically for adhesive bonding; seldom can an assembly designed for another method of fastening be successfully bonded without being modified. Adhesively bonded joints must be stressed in their strongest directions – in tension, shear, and compression – and load must be minimized in the peel and cleavage directions.

- Temperatures: The adhesive and the substrate can become brittle due to low temperatures or may melt or decompose under conditions of extreme heat. Heat sensitivity of adhesives is an important consideration in their selection and can be one of their most severe limitations. While some can withstand temperatures as high as 700 °F, most are limited to service under 200 °F. Most high-temperature adhesives require an oven cure, although some cure at room temperature. If an adhesive is not formulated for high-temperature service, its strength drops considerably in such environments. Low temperatures, on the other hand, make many adhesives brittle and stress joints internally.
- Harsh exposure: Consider water and humidity, as well as exposure to solvents and such.

Other items include flexibility, aging stability, and aesthetic questions, such as what are the joint's flexibility requirements? And its life span? What color is required and what level of gloss?

- Cost factors relate the cost of adhesive bonding to other values in the manufacturing process. If several adhesive systems meet the requirements for an application but significantly differ in price, more detailed analysis could determine an actual bonding cost per unit. Criteria involved include waste, process speed, rejects/failures, packaging, reliability, availability, and service.

- Waste: Waste increases the cost per bond and the impact on the machinery. Cleaner application characteristics lead to less downtime and increased use of production equipment.

- Process speed: If one adhesive provides faster production speeds than the others, that value should be included in the cost/value ratio.

- Rejects and field failures: Rejects are a fact of life, and usually an acceptable level of rejection has been established. If an adhesive system offers performance properties that reduce reject rates, higher initial costs can be overcome and profits may increase.

By figuring the costs surrounding packaging considerations, product consistency, batch-to-batch reliability, product availability and the strength of the supplier's backup service – the total cost of each contending adhesive presents itself.

Adhesive application: Adhesives can be applied with any type of liquid-handling tool, such as a brush, spatula, trowel, dip, spray, curtain, flow gun, or flow brush. However, production-line conditions generally require the use of automatic or semiautomatic dispensing equipment which can apply dots, bands, or beads. Most serious designers will not consider an adhesive without also considering a suitable application method. Figure 6.35 shows a typical process used in adhesive bonding [40]. Figure 6.36 shows the typical failure modes of adhesive joints [41].

Dip coating and spraying can also be used for flat parts. These are especially suitable for contoured parts. Brushing is widely used to apply liquid and thin-paste adhesives: Equipment is simple, waste is minimal, and limited areas of contoured shapes can be coated without masking. However, high production rates and uniform adhesive thicknesses are difficult to achieve.

6.3.3 Surface Pretreatments

Because adhesives bond to surfaces, the actual surface condition is most important. As a rule, the surfaces to be joined should be as clean and dry as practically possible. Surface pretreatment will, therefore, normally be necessary if optimum performance of the adhesively bonded joint is aspired. Proper surface preparation generally improves both adhesive bonding and paint adhesion. On the other hand, alloy differences or even different materials are of less concern if the appropriate surface treatment methods are used and the applied adhesive bonds to both sides of the joint.

Thermal oxidation for bottom insulation 1µm

SU-8 spin coating/patterning 2-8 µm

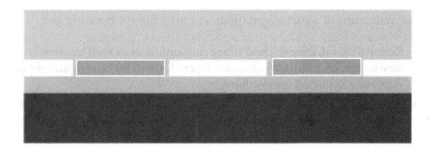

Bonding with counter wafer

FIGURE 6.35
Adhesive bonding process [40].

Surfaces are likely to be contaminated with materials that could adversely affect joint performance. Therefore, impurities (i.e., dust, dirt, oil, grease, fat, or water) and the inactive adsorption layer created by foreign molecules (i.e., water and gases) are generally first removed. Care must be taken to avoid contaminating the surfaces during or after pretreatment. Contamination may be caused, for example, by part-handling ("finger-marking") or by other working processes taking place in the bonding area, e.g., oil vapors from machinery, metal dust from abrasive processes or vapors from spraying operations (paint, mold release agents, etc.)

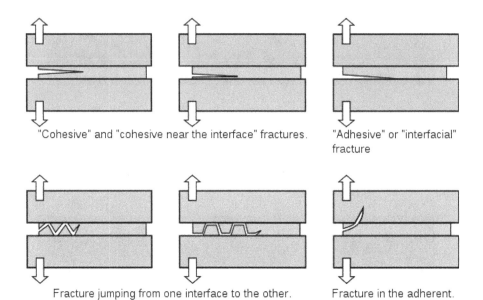

"Cohesive" and "cohesive near the interface" fractures.   "Adhesive" or "interfacial" fracture

Fracture jumping from one interface to the other.   Fracture in the adherent.

FIGURE 6.36
Failure of the adhesive joint can occur in different locations [41].

The main aims of a surface pretreatment for adhesive bonding are:

- Removal of oil, greases, and other contaminants as well as other surface layers, including weak oxide layers formed by heat treatment or exposure to humid atmosphere
- Protect the substrate surface prior to bonding and maximize the degree of intimate molecular contact between the adhesive and the substrate surface
- Generation of stable surface topography for optimum mechanical interlocking
- Improve durability and corrosion resistance of the adhesively bonded joint; promote the formation of intrinsic adhesion forces that exhibit both resistance to environmental attack by moisture and chemical stability over a wide pH range

6.3.4 Surface Layers of Metallic Parts

Most surfaces are covered with layers of contaminants, for example, dust, oils, grease, and oxide. These contaminants adversely affect the adhesion and wetting of the adhesive and if they are not removed properly before applying the adhesive, it can result in bond failure. The surface layers of metallic parts are as follows (Figure 6.37):

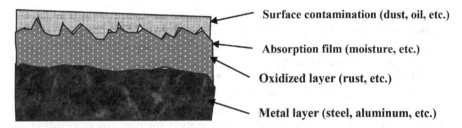

Surface contamination (dust, oil, etc.)

Absorption film (moisture, etc.)

Oxidized layer (rust, etc.)

Metal layer (steel, aluminum, etc.)

FIGURE 6.37
Typical surface layers on a metallic part (adapted from ref [42]).

- Metal surface: This is the base structure of metal. This is the part where deformations and cracks occur due to shear or other forces.

- Oxide layer: This layer is formed due to the adsorption on the base metal surface.

- Absorption film: This layer forms due to the absorption of moisture from the environment.

- Contamination: This layer is formed when the dust, oil and other contaminants settle on the surface.

6.3.5 Methods of Surface Pretreatment

Following the preliminary surface preparation, it may be necessary to pre-treat the surface using one of the varieties of mechanical, chemical, or physical methods.

- **Mechanical Methods [43]**

Mechanical methods involve the use of handheld sandpaper or hand-cleaning tools, such as wire brushes and scrapers. Such instruments are convenient to remove rust, scale, paint, and weld splatter. But they are too slow to use in large areas.

One drawback of the abrasion process is that it causes particles of debris to accumulate on the abraded surface. These particles come from the abrasive, the surface contaminants, and the surface material itself. All such particles must be removed before the adhesive is applied. This may be accomplished with a clean cloth or brush, or with filtered compressed air. After the abrasion debris has been removed, it is usual to give another solvent clean before bonding. A solvent-moistened cloth is convenient for this, but as the cloth will become contaminated during this operation, it should be renewed frequently.

Mechanical cleaning also includes a number of much faster-abrading methods, such as sandblasting, tumbling, and abrading with power tools.

- **Chemical Methods [43]**

In chemical treatments that alter the surface of the adherend, the part is dipped into a chemically active solution. This solution either dissolves part of the surface or transforms it, making it more chemically active and thus more receptive to adhesive bonding.

Acid etching involves immersing a metal substrate in an aqueous acid solution to remove a loose layer of oxide from its surface. The particular acid used depends upon the metal and type of oxide being treated. In many cases, acid etching may provide enough surface preparation for bonding depending, of course, upon the degree of adhesion desired. Acid etching can also be effectively used with certain plastics; for example, chromic acid is used to treat polyolefins.

Anodization involves the electrochemical modification of the surface. The process deposits a porous and stable oxide layer on top of the oxide layer formed after etching of the substrate.

- **Physical Methods** [43]

These are techniques where the surface is cleaned and chemically modified by exposure to highly energetic charges or other ionic species. The most common methods are flame treatment, corona discharge, and plasma. These pretreatment methods have been applied to metals and, in particular, composites and plastics.

- Flame treatment of the substrate surface for just a few seconds with an oxygen-containing (blue) propane or acetylene gas flame leads to the incorporation of oxygen-containing groups at the surface. This improves the wetting properties and, hence, the adhesion. Flame treatment is used almost exclusively for polyethylene and polypropylene substrates. The effect of the pretreatment subsides within a short time so that flame-treated substrates must be bonded immediately.

- Reference [43] shows the photograph of a corona discharge, which is essentially plasma generated in air at atmospheric pressure by applying a high frequency and high voltage between two electrodes. It contains a number of energetic species that can clean and introduce polar groups, mostly oxidation products, at the substrate surface. The corona discharge may also lead to crosslinking of the polymer surface. The pretreatment effects are shortlived. So. bonding should be carried out immediately. It is used mainly for polyolefin films and is capable of high processing speeds.

- Plasma cleaning is another method which is usually generated in a low-pressure chamber and so is best suited to batch processing. Commercial units of various sizes are available. The advantage of this method is that it allows treatment of substrates by plasmas of gasses other than oxygen, for example, argon, ammonia, or nitrogen. Plasmas created from inert gases are generally used to clean substrate surfaces.

6.3.6 Surface Pretreatments for Aluminum

The aluminum surface is a complex transition zone between the bulk of the alloy and the environment. A thin natural aluminum oxide layer lies on top of a subsurface layer with different chemical and microstructural properties than the bulk material. In rolled aluminum products, the disturbed ("deformed") surface layer generally has a thickness of a few 100 nm. It contains rolled-in oxide particles from the high forces used during hot and cold rolling; it has elongated and smaller grains than the bulk and it can contain various intermetallic particles different in chemistry and concentration to the bulk.

The natural aluminum oxide film has a thickness of 5–20 nm and follows exactly the material surface topography. In principle, this oxide film would present an ideal basis for adhesive bonding. In fact, a high initial joint strength can be often obtained without any pretreatment or by simple degreasing of the aluminum surface before adhesive bonding. But in order to maintain the integrity of bonded joints, in particular, in humid environments, some form of surface pretreatment is always necessary, specifically if the joints are subjected to tensile stresses.

The underlying reason is that the natural oxide film is not perfect. It may show thickness variations and exhibit defects, such as pores and fine cracks; the aluminum oxide can be disturbed locally (e.g., where intermetallic particles are present in the adjacent aluminum metal) and it can contain surface contaminants (e.g., residues of rolling oils). Furthermore, depending on the thermal history (temperature and time), solute alloying and impurity elements can diffuse from the matrix into the surface oxide layer, modifying the structure and composition of the surface oxide film. Surface enrichment and the formation of a heterogenous surface oxide layer are particularly pronounced for elements, such as Mg, Li, Na, Be, and Ca. These effects degrade the characteristics of the surface oxide layer, i.e., make it more hygroscopic and less corrosion-resistant. Therefore, the natural, inhomogenous aluminum oxide surface film is often removed and replaced by a properly controlled, new, homogenous surface film.

Depending on the specific aluminum product (sheet, extrusion, casting, etc.), the applied surface pretreatments in preparation for adhesive bonding may be somewhat different; however, the individual steps are essentially the same.

The three processes in pretreatment of aluminum are as follows:

- Clean
- Etch
- Seal (followed by dry-off)
- Cleaning and removal of all-surface contamination that may inhibit adhesion or application of pretreatment or conversation coating are critical and normally accomplished with surface-active agents (surfactants) built into a complex cleaner.

- Aluminum, such as many nonferrous and plastic substrates, requires a mild microscopic surface etch prior to coating to promote improved adhesion. This can be accomplished simply and safely, with the right chemistry additives in your pretreatment chemistry. The industry-standard chemistries used to microscopically etch aluminum have been primarily acids, such as phosphoric, sulphuric, or chromic acids, many times used in combination, somewhere in the pretreatment process with a fluoride-containing additive to promote the etch. Hydrofluoric acid was used in the past to etch and brighten aluminum. Hydrofluoric acid should be or literally could be your last choice to utilize. This is due to the extremely hazardous and potentially deadly handling concerns and extremely low human threshold limits. Consultation with a pretreatment representative for the best etchant and/or reviewing conversion chemistry available for specific process requirements is the safest way to follow.

- The seal is many times the last process applied prior to a low-conductivity, deionized, distilled or reverse osmosis-conditioned water rinse. The purpose of the seal process is to completely passivate and seal any microscopic voids on the surface from the environment prior to coating. This helps additionally improve adhesion and inhibit corrosion. The seal process can be accomplished in many ways with today's technological advancements. In some of the latest pretreatment technologies, the seal and passivation steps are accomplished in a single step during the conversion-coating application process. The chemistry can be comprised of Hexavalent Chrome (which is a known carcinogen in the hexavalent state), trivalent chrome (which is thought to be a safe alternative to the hexavalent chrome), nonchromic seal, such as molybdate and other organic, inorganic, and nanopolymer products are commercially available.

6.4 Bonding Strength

Joint design incorporating adhesives requires specific attention because of the large property differences between the adhesive and the materials being bonded. In order to correctly understand the effect of different adhesives on a bonded joint, the bonded joint must be considered as an independent structural element in a composite structure. It has different mechanical properties which can be influenced in a different manner by temperature and other environmental conditions. Since the durability of the various adhesives in different environments is generally known, the selection of an appropriate adhesive presents normally little problems.

Low-strength bonded aluminum joints are often a boundary layer problem, i.e., the result of undesired effects between adhesive and aluminum oxide. Water, either in its liquid or vapour phase, is the most common and generally most severe environmental stress factor. Different adhesives can differently interact with the aluminum surface in the boundary layer as a result of the electrolytes that can form in the presence of water. The effect on the boundary layer is even more negative if the water contains salts.

6.4.1 Interdependence of Material

The peculiar behavior of the strength of adhesively joint metals is a result of the fact that the joint system is not homogenous but consists instead of a composite system in which the resulting properties are a combination of the individual properties of the parts to be jointed: the adhesive layer and the interface layers.

The specific properties of the adhesive joint are a result of the optimal strengths obtained due to the geometrical and material design as shown in Figure 6.38 (adapted from ref [42]). This is an iterative process until all constraints are optimally satisfied.

6.4.2 Loading Factors

The overall performance of an adhesive metal joint is characterized by the measure in which it is able to withstand loads without any appreciable change in its original strength values. Figure 6.39 shows the types of stresses induced in typical adhesive joints.

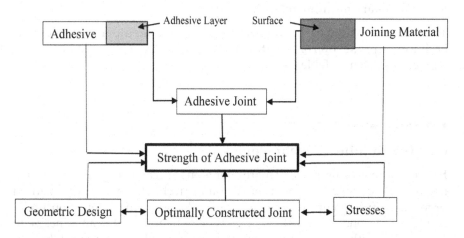

FIGURE 6.38
Typical surface layers on a metallic part (redrawn and adapted from ref [42]).

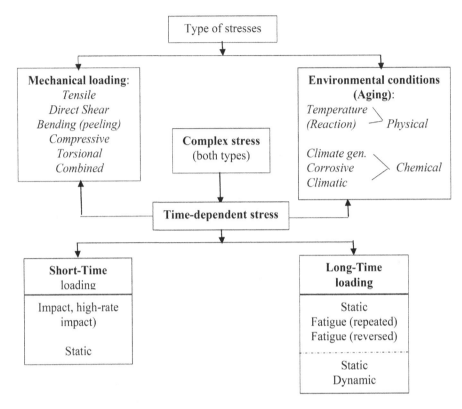

FIGURE 6.39
Types of stresses on adhesive joints (redrawn and adapted from ref [42]).

6.4.3 Summary of Influencing Parameters

The combined action of the influencing factors and their parameters are the basis for the production of an optimal adhesive joint and govern its attainable strength as given in Table 6.1:

6.5 Hybrid Joining Technologies

Hybrid joining is where two or more operations are carried out either simultaneously or sequentially. These types of joining technologies lead to enhanced properties of the joint due to a synergistic load-bearing interaction under service conditions. The most common type of hybrid joint is an adhesive in conjunction with a point joint, such as a mechanical fastener (rivet, bolt, or screw) or a spot weld. Hybrid joining is mainly used for joining sheet materials but there are also applications involving extrusions and thin castings.

TABLE 6.1

Parameters Influencing the Strength of Adhesive Joints in Metals (Raghu's Notes; Adapted from Ref [42])

Adhesive Layer	Joining Material	Geometric Design	Type of Stress Induced
Elastic modulus, E_A	Elastic modulus, E_M	Overlap length, L_o	Mechanical
Modulus of rigidity, G	Ultimate tensile strength, S_U	Overlap width, b	Physical
Poisson's ratio, v	Tensile yield strength, S_Y	Thickness of joining part, t_J	Chemical
Stress-shearing behavior	0.2% Offset method for S_Y	Thickness of adhesive layer, t_A	Combined mechanical, physical, chemical, and time-dependent
–	Poisson's contraction	–	–

6.5.1 Application Criteria

The four main types of joining typically used in hybrid joining technologies are some combination of the following:

- Adhesive joining
- Spot welding
- Clinching
- Riveting

Each different joining technology has characteristics that make it a good choice for some applications and poor for others. Table 6.2 below shows the characteristics of each of the above joining technologies. It can be seen that a combination of two of the joining processes can give more characteristics to the joint that is not possible by just using one joining technology.

The classification index of a joint is a value for the functional characteristics relative to the corresponding base material of the corresponding material combination. The joining technology used in light construction should be such that the classification index of the joint optimally approaches the limiting value of 1 (Equation 6.2). At increased stresses, the principles of force actions, the reduction of disadvantageous notch effects and the consistency of material properties and the joint reliability become increasingly important. A consequence of the above-mentioned issues is that there is an increasing tendency to use aluminum for highly stressed constructions.

$$\frac{Functional\ Properties\ of\ Joint}{Functional\ Properties\ of\ Base\ Material} \rightarrow 1 \qquad (6.2)$$

TABLE 6.2

Comparison of Functional Characteristics of Different Fastening Technologies
(Raghu Retyped for His Notes; Adapted from Ref [42])

Joining Technology	Adhesive Joining	Spot Welding	Clinching	Riveting
Functional Characteristics	Load carrying	Load carrying	Load carrying	Load carrying
	Fixing	Fixing	Fixing	Fixing
	Sealing	Electrically conductive	Electrically conductive	Electrically conductive
	Isolating	Relatively low fatigue strength	Relatively low fatigue strength	Relatively low fatigue strength
	Damping			
	Equalizing			
	Good fatigue strength			

A combination of joint types can be used, among others, to take advantage of the specific material properties of aluminum. Therefore, making it possible to optimize the joint quality, allow certain materials or material combinations to be joined, and/or to simplify the fastening process. A combination of different joint types can be used either to improve the static and dynamic properties of the joint or to guarantee leakproof joints (Figure 6.40).

Depending on the load-carrying capacity and the design of light construction, the types of joints used may be of the material-locking kind or of the force- or shape-locking kind. The choice of elementary joints used

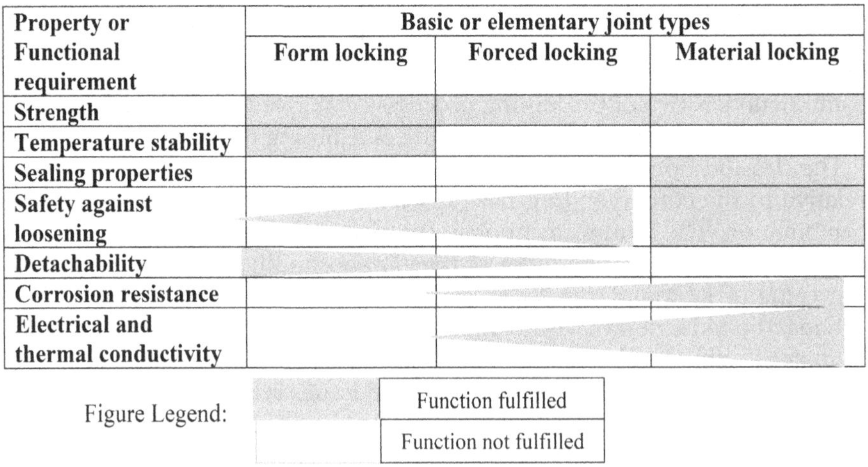

Property or Functional requirement	Basic or elementary joint types		
	Form locking	Forced locking	Material locking
Strength			
Temperature stability			
Sealing properties			
Safety against loosening			
Detachability			
Corrosion resistance			
Electrical and thermal conductivity			

Figure Legend:	Function fulfilled
	Function not fulfilled

FIGURE 6.40
Application of elementary joints (redrawn and adapted from ref [42]).

TABLE 6.3

Advantages and Disadvantages of One-Sided Overlapped Adhesive Joints (Adapted from Ref [42])

One-Sided Overlapped Adhesive Joints	
Advantages	Disadvantages
Force transmission over large areas	Sensitive to peeling forces
No thermal influence of material microstructure	Aging problems
Suitable for different types of materials including nonmetals	Limited warm strength

in combination is based on the principle of elimination. According to the Principle of Elimination, those particular combinations of joints, which do not fulfill any one out of a required list of criteria mentioned in Figure 6.40, are either not considered or eliminated.

The specific disadvantages of adhesive joining can be compensated for by using a combination of elementary joints. Examples of this can be found in the automotive and aerospace industries (Table 6.3).

In the automotive industry, the combination of adhesive joining and folding can be used for fastening car body parts [42]. If applied properly, the folded and adhesively joined parts possess the combined advantages of both fastening technologies. The main advantage of folded joints is that these can be loaded immediately (a rational production is possible). The surfaces of the folded joints remain smooth and clean. The additional use of adhesives in the fold leads to leakproof joints. Additionally, at the same time, the adhesive used improves the damping characteristics of the whole aluminum construction.

Joints consisting of a combination of both adhesive joining and locally active fastening methods mostly use spot welding or mechanical joining methods, such as clinching or riveting. In the automotive industry, spot welding is used mainly to shorten production time in spite of the long hardening times of adhesives. The combination of adhesive and riveted joints is used frequently in the aerospace industry primarily for parts subject to dynamic loading. During the clinching process, aluminum-shaped sheet components and profiles are joined together according to the quasi form-locking principle simply through the action of local plastic material deformation without the use of auxiliary parts or thermally influencing the microstructure. Material spray, which occurs as a joint defect due to the high material pressure in the spot-welded region, thereby reducing the joint strength, is absent in joint combinations of clinching an adhesive joining. Some adhesive-sealed joints are prone to deleterious aging, especially if exposed to an industrial atmosphere, water, solvents, or other aggressive chemicals. Examples of combined adhesive and sealed joints used in aircraft construction illustrate how this deleterious effect can be reduced by the judicious choice of adhesive and sealant, the longtime testing of joints, a specific design as well as application of protective layers on the surface of the adhesive and sealed joints.

6.5.2 Production Considerations

Joint combinations consisting of more than one elementary joint can be fabricated by producing the individual joints simultaneously or one after the other. The order that the joints are "assembled" can influence the properties of the joint. For example, if it is a single process, the joint properties can be reached prior to use or during the use of that joint. However, if the number of processes is more than one, then some of the joining processes can be performed simultaneously prior or during the use of the joint. Other types of joining processes have to be undertaken consecutively, i.e., prior to or during use [42].

In principle, three variants are possible for the consecutive production of adhesive joining and mechanical joining. In the "capillary method", a mechanical joint is first prepared and then an adhesive of low viscosity is brought into the joint crevice. In this method (which is characterized by a clear separation of the parts to be joined), the adhesive serves generally as a sealant and/or an inhibitor for corrosion. The adhesive joining of sheet and profile parts followed by a mechanical fastening process, the latter being used to improve the peeling strength of the joint, has till now played only a secondary role. Another fastening process has been found to be industrially most suitable for making combined joints, especially for mass production. In this process, the adhesive is first applied to the parts and then, before the adhesive hardens, a mechanical joining process is applied which goes through the unhardened adhesive. The hardening then follows as usual depending on the type of adhesive used. A closer look at the individual process steps involved in the production of the combined adhesive-clinch joint shows that the combined adhesive-mechanical joining process can be easily integrated into the mass production of thin sheet constructions. Figure 6.41 shows the block diagram of this production process.

In adhesive joints combined with mechanical joining, the former is the main joining process. The mechanical joining serves as a positioning (fixing) help and helps the adhesive joint to withstand peeling forces and long-time static forces. For mechanical joining combined with adhesive joining, clinching is the most common joining process. The applied adhesive serves primarily as a sealant, corrosion protection, and/or damping material and relieves the joint in regions where the force lines line outside the joining point. Clinching without local incision can be used as the mechanical fastening partner for adhesive joining using pasty adhesives. On the other hand, clinching with the local incision is used together with adhesive foil and

FIGURE 6.41
Technical operations of adhesive and mechanical combination joining.

bands. The geometry of the joint element of combined clinched fastenings illustrates that in spite of the presence of adhesives, an optimal form of the clinch joint is created, assuming, of course, that appropriate process technology is applied. The advantage of using clinching instead of spot welding in combination with adhesive joining is that the former allows the use of not only fluid and pasty adhesives but also of adhesive foils and bands. Solid adhesives have special advantages, both as far as process technology as well as health aspects are concerned.

The properties of the adhesive used are the main deciding criteria for the application. Under quasi-static, dynamic, and impact loading, the load-carrying capacity of, for example, clinched adhesive joints compares well with that of spot-welded joints. Under the action of dynamic loads, the geometric notch effect of the "point-formed" joint element has a deleterious effect on the load-carrying capacity of the combined joint. Joints made of ductile adhesives having high deformability and low strength behave differently. Here again, depending on the loading, the joint strength is determined mainly by the clinched joint. By giving proper consideration to factors, such as property profile and processing properties of the adhesive, it is basically possible to design combined joints with properties which are a combination of the individual properties of the joints.

In connection with combined fastenings of mechanical and adhesive joints, the aspect of aging of the combined joint plays a central role. It has been found that corrosion increases the strength of riveted and clinched joints of aluminum sheets. This is due to the fact that corrosion products of aluminum occupy a larger volume than the uncorroded aluminum material, therefore increasing the strain on the joint causing the force-locking component to increase. Although the mechanical fastening process can cause aluminum sheets to be pulled apart in the point vicinity, the supporting action of the mechanical joint in the combined mechanical-adhesive joint greatly reduces the decrease in joint strength caused by aging. Mechanical fastening methods, used alone or in combination with adhesive joining, improve the standard of quality as far as damping of noise and vibrations, pressure tightness, and corrosion protection are concerned, making this an interesting proposal for highly stressed aluminum constructions. Table 6.4 shows a comparison of properties from different joining technologies and their sensitivity to the environment:

6.6 Joining Dissimilar Metals

As aluminum alloys are more frequently applied in the automotive industry, the joining of aluminum not only to itself, but also to other materials is increasingly important. When joining aluminum to other materials, three different tasks can be differentiated:

- Joining aluminum to compatible metals with some degree of solubility in each other

TABLE 6.4

Comparison of Properties From Different Joining Technologies (Adapted from Ref [42])

Joining Technology	Adhesive Joining	Spot Welding	Clinching	Riveting
Joint	Large surface	Local	Local	Local
	Plane surface	Uneven surface	Deformed surface	Damaged surface
Properties	Very sensitive to the environment	Conditionally sensitive to the environment	Conditionally sensitive to the environment	Sensitive to the environment
	Depend on direction	Depend on direction	Depend on direction	Depend on direction

- Joining aluminum to incompatible metals with little or no solubility in each other
- Joining aluminum to different types of material, such as plastics, composites, and ceramics.

The following processes are applicable when joining aluminum to other metals:

- Fusion welding
- Arc welding
- Beam welding
- Resistance welding
- Solid-state joining processes
- Brazing
- Soldering
- Mechanical joining processes
- Adhesive bonding

However, when aluminum must be joined to other types of materials, such as plastics composites and ceramics, fusion welding methods cannot be applied.

6.6.1 General Issues and Limitations

A number of factors must be taken into consideration when designing a dissimilar material joint including:

- Material combination and performance requirements
- Joint design and material thickness

- Thermal expansion and contraction mismatch during joining and service
- Potential galvanic corrosion problems during service
- Fixture requirements and constraints regarding joining stresses

Depending on the specific joining process, additional factors may have to be considered as well. In the case of fusion welding, differences in melting temperature, the formation of brittle intermetallic compounds during joining (leading to brittle joints), the heating and cooling rates (effect the microstructure of the joint), the need for pre- and postheating (to minimize stresses during welding and cooling), and the need for composite transition materials or special filler materials during joining, etc., are all important considerations.

6.6.1.1 Metallurgical Limitations

Aluminum is, in general, difficult to weld to other materials. For this reason, joining aluminum to other metals has been mainly done using other joining methods (particularly mechanical joining and adhesive bonding) instead of fusion welding. However, new developments have led to a renewed interest in fusion welding processes. When joining aluminum to other metals, the major difficulty is that at high temperatures, and, in particular, in the presence of a liquid phase, brittle intermetallic compounds are formed at the interface, which results in poor joint characteristics. As an example, fusion welding of aluminum to steel leads to the formation of particles of intermetallic phases, such as $FeAl_2$ and Fe_2Al_5. Brittle intermetallic phases are also formed when "fair weldable" metals, such as copper, magnesium, or titanium, are directly fusion-welded to aluminum.

6.6.1.2 Galvanic Corrosion

This topic was discussed before in Chapter 1. Galvanic corrosion is an electrochemical process in which one metal corrodes preferentially to the other. Both metals must be in electrical contact and in the presence of an electrolyte. Dissimilar electrically conductive materials have different electrode potentials and when two or more come into contact with an electrolyte, one material can act as an anode and the other can act as a cathode. The difference in electrode potential between the dissimilar metals is the driving force for an accelerated corrosion attack on the anode member of the galvanic couple. The anode metal dissolves in the electrolyte, and the corrosion products deposit on the cathode.

The electrochemical reaction that takes place is an oxidation-reduction reaction (redox). An oxidation reaction is the removal of one or more electrons from an atom, ion, or molecule; this takes place at the anode. A reduction reaction is the addition of one or more electrons to an atom, ion, or molecule; this takes place at the cathode. When a redox reaction takes place, electrons

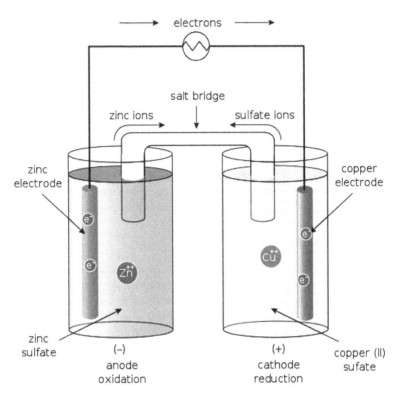

FIGURE 6.42
Schematic of zinc-copper electrodes of a galvanic cell [44].

flow from one metal to another causing a current flow and, consequently, corrosion. The corrosion is concentrated at the metal that is less noble (called the anode) which can be determined using the galvanic series. The galvanic series ranks metals and alloys as to their relative electrochemical reactivity in seawater. The galvanic series is practical and essential for determining the effects of galvanic corrosion (Figure 6.42). Similar reactions take place between Fe-Cu cells.

The galvanic series is the guiding principle for determining the effects of galvanic corrosion. It can be seen in the figure above (Figure 6.43) that the more inert metals (cathodic metals) are towards the bottom and the more reactive metals (anodic metals) are towards the top. The rule of thumb is that metals that have 0.2 V of voltage difference are at risk for galvanic corrosion. Additionally, the greater the voltage difference (further apart the metals are on the table), the greater is the corrosion risk. The figure below illustrates the corrosion risk for two metals. Green indicates that corrosion risk is low or negligible, and red indicates that corrosion is a risk for that material combination.

No.	MATERIAL	VOLTAGE RANGE	RELATIVE POSITION
1	Magnesium	−1.60 to −1.67	
2	Zinc	−1.00 to −1.07	
3	Beryllium	−0.93 to −0.98	
4	Aluminum Alloys	−0.76 to −0.99	
5	Cadmium	−0.66 to −0.71	
6	Mild Steel	−0.58 to −0.71	
7	Cast Iron	−0.58 to −0.71	
8	Low Alloy Steel	−0.56 to −0.64	
9	Austenitic Cast Iron	−0.41 to −0.54	
10	Aluminum Bronze	−0.31 to −0.42	
11	Brass (Naval, Yellow, Red)	−0.31 to −0.40	
12	Tin	−0.31 to −0.34	
13	Copper	−0.31 to −0.40	
14	50/50 Lead/Tin Solder	−0.29 to −0.37	
15	Admiralty Brass	−0.24 to −0.37	
16	Aluminum Brass	−0.24 to −0.37	
17	Manganese Bronze	−0.24 to −0.34	
18	Silicon Bronze	−0.24 to −0.30	
19	Stainless Steel (410, 416)	−0.24 to −0.37 (−0.45 to −0.57)	
20	Nickel Silver	−0.24 to −0.30	
21	90/10 Copper/Nickel	−0.19 to −0.27	
22	80/20 Copper/Nickel	−0.19 to −0.24	
23	Stainless Steel (430)	−0.20 to −0.30 (−0.45 to −0.57)	
24	Lead	−0.17 to −0.27	
25	70/30 Copper Nickel	−0.14 to −0.25	
26	Nickel Aluminum Bronze	−0.12 to −0.25	
27	Nickel Chromium Alloy 600	−0.09 to −0.15 (−0.35 to −0.48)	
28	Nickel 200	−0.09 to −0.20	
29	Silver	−0.09 to −0.15	
30	Stainless Steel (302, 304, 321, 347)	−0.05 to −0.13 (−0.45 to −0.57)	
31	Nickel Copper Alloys (400, K500)	−0.02 to −0.13	
32	Stainless Steel (316, 317)	0.00 to −0.10 (−0.35 to −0.45)	
33	Alloy 20 Stainless Steel	0.04 to −0.12	
34	Nickel Iron Chromium Alloy 825	0.02 to −0.10	
35	Titanium	0.04 to −0.12	
36	Gold	0.20 to 0.07	
37	Platinum	0.20 to 0.07	
38	Graphite	0.36 to 0.19	

(Left margin, top to bottom: LEAST NOBAL (ANODIC) ↑ ... MOST NOBAL (CATHODIC) ↓)

FIGURE 6.43
Galvanic series [45].

Galvanic corrosion is most severe near the junction between the two dissimilar metals and the severity of the corrosion decreases further away from the junction. Like all types of corrosion, the galvanic corrosion rate is significantly affected by the area exposed to the electrolyte (Figure 6.44). However, for galvanic corrosion, the area of the metals (anode and cathode) in contact with the electrolyte can be different and this can drastically affect the corrosion rate. Specifically, the ratio of the cathodic area to the anodic area is a good indicator of corrosion rate. If the idea is to moderate the effects of corrosion, it is desirable to have a small ratio of cathode area to the anode area. For

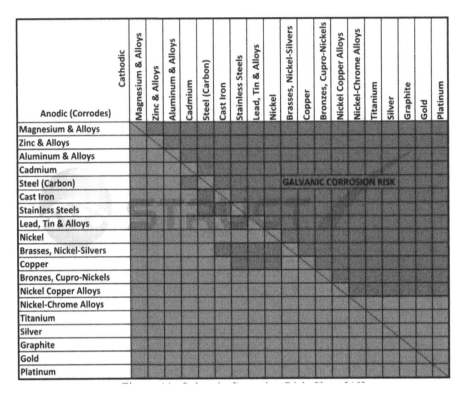

FIGURE 6.44
Galvanic corrosion risk chart [45].

this reason, only one of the dissimilar metals needs to be coated for corrosion protection and the metal that is nobler (cathode) or corrosion-resistant should be coated (Figure 6.45).

The severity of any corrosive action can be mitigated or prevented by taking the appropriate measures. Some of the common means of corrosion prevention are:

- Proper material selection
- Environmental modification
- Plating/protective coatings
- Cathodic protection
- Passivity

Adding a protective layer or coating to a material can prevent corrosion as long as the coating stays intact. As discussed above, pitting is common in situations where the protective coating has chipped off. Painting, electroplating, chemical plating, and mechanical plating are common processes that

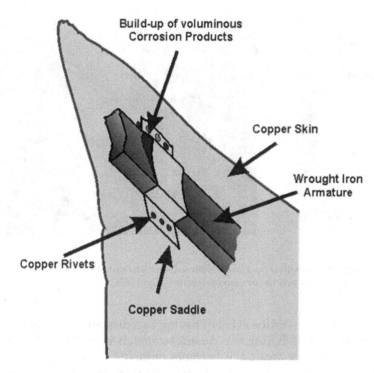

FIGURE 6.45
Galvanic corrosion in the Statue of Liberty [46].

coat the material and help prevent the effects of corrosion. The painting covers the material but is susceptible to chipping. This is common in the automobile industry. Car bodies' rust (a form of corrosion) and the paint protect the metal from encountering a corrosive environment (atmosphere, water, etc.). Electroplating is where a thin layer of metal is deposited on the part in an electrolytic bath. This thin layer protects the part from the potentially corrosive environment. Steel parts are often electroplated with nickel, tin, or chromium. Figure 6.46 shows the principle of this process.

Mechanical plating is where the part is tumbled with a metal powder. The tumbling action causes the metal powder to be cold-welded to the part. This process is common for fasteners, such as screws, nails, and washers. Mechanical plating is advantageous because this method avoids hydrogen embrittlement. Hydrogen embrittlement is when a metal becomes brittle due to the diffusion of hydrogen into the metal (common in high-strength steel). Hydrogen embrittlement is a concern with electroplating.

Environment modification and proper material selection are two very important ways of preventing corrosion. In some cases, the environment can be modified to keep water or other corrosive materials out of contact with the metals. In other cases, different fastening techniques can be used

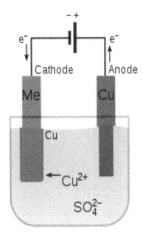

FIGURE 6.46
Electroplating a metal with copper in a copper sulfate bath (left); Aluminum anodes mounted on a steel-jacketed structure as corrosion protection (right) [47].

to eliminate crevices or the event of having two dissimilar metals in contact with one another. If joining two dissimilar metals is unavoidable, trying to get favorable surface area ratios becomes more important. A smaller anode will corrode more rapidly than a larger one; therefore, make the anode as small as possible. Lastly, if two dissimilar metals are required to be next to each other, insulating them electrically will help diminish corrosion. When the environmental modification is not an option, material selection becomes increasingly important. Selecting materials that resist corrosion is one way of preventing the corrosion. However, these materials are typically very expensive or not suitable for mechanical applications (gold, platinum, and titanium are good examples). In this case, it is advantageous to select materials that not only can perform the desired function but are also similar to each other. Looking at the galvanic series above and choosing materials that are more similar can mitigate the effects of corrosion. Lastly, choosing materials that passivate under normal operating conditions will also help prevent corrosion. Passivation is where normally active materials (that will corrode) lose their reactivity. This is due to the formation of a very thin oxide layer on the surface of a material that stops further reactions. Passivation is responsible for the high corrosion resistance of stainless steel (nickel and chromium oxides).

The final way of preventing corrosion is by cathodic protection. Figure 6.47 shows a source of DC electric current used to help drive the protective electromechanical reaction (on the left). This method requires supplying electrons to the metal to be protected. This reverses the current flow essentially making the material aimed to be protected an artificial cathode. However, the protected metal must be electrically connected to another metal that is

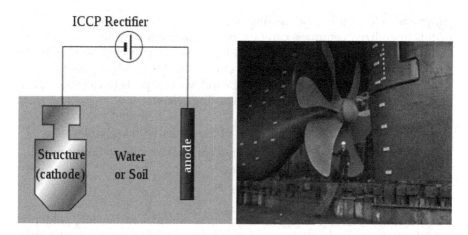

FIGURE 6.47
Sketch of a simple impressed current cathodic protection system (left); White patches visible on the ship's hull are zinc block sacrificial anodes (right) [48].

more reactive; this metal is called a sacrificial anode. Magnesium and zinc are common materials for sacrificial anodes because they are among the least noble metals. Sacrificial anodes are commonly used in outboard engines so that the aluminum gearcase housing does not corrode in the water.

6.6.1.3 Thermal Expansion

Thermal expansion is the tendency of matter to undergo a volume change in response to a temperature change. For solids, the main concern is the change along a length over some area. The coefficient of thermal expansion is a material-specific parameter and generally varies with temperature. However, common engineering solids usually have coefficients of thermal expansion that do not vary significantly over the range of temperatures where they are designed to be used, thus practical calculators can be based on an average value of the coefficient of expansion. When the applied joining technique involves significant temperature changes, thermal expansion effects must be considered such as the joining operation of dissimilar materials. Additionally, for body-in-white applications, however, the most important post-joining effect is the lacquer bake hardening process which takes place at temperatures up to about 180 °C.

6.6.2 Joining Aluminum to Other Metals

6.6.2.1 Brazing and Soldering

As discussed previously, brazing and soldering have a significant advantage over other molten metal joining techniques. The formation of brittle intermetallic compounds can be significantly inhibited through the use of brazing alloys and solders with low melting points. Dissimilar metals and even non-metals (metalized ceramics) can be joined to aluminum in this manner. For joining ceramics to metals, thin metal layers are usually deposited onto the ceramic part prior to brazing in order to facilitate the bonding process.

When soldering dissimilar metals, the following aspects have to be considered when selecting an appropriate soldering system:

- The compositional compatibility of the solder with both interfaces
- The differences in the coefficient of thermal expansion between the two materials
- The differences in melting points

Since aluminum has a high coefficient of thermal expansion, soldering (which can be carried out at much lower temperatures than brazing) may be the preferred solution in many applications.

6.6.2.2 Friction Welding

The principle of friction welding is that the heat required for the welding process is produced by friction. The main friction welding processes are the following:

- Inertia friction welding
- Linear friction welding
- Stir friction welding

For inertia friction welding, a flywheel accelerates and the workpieces are brought into contact with each other. Once the workpieces are in contact, an axial force is applied (forcing the two materials together). This force causes the flywheel to slow down and eventually stop. The main limitation of this process is that it is restricted to parts where angular orientation is not important. Figure 6.48 shows the principle:

Linear friction welding has been covered earlier in this chapter (Figure 6.48). For completeness, it is briefly covered in this section again. The photograph in Figure 6.49 shows a reciprocating relative linear motion between the faces of two parts, which creates friction. Usually, one of the parts is fixed and the other part linearly translated. This process can be used in situations where orientation is important. This process is often used when joining dissimilar metals, such as aluminum and steel. For more details about the rotary and linear friction welding principles, one can refer to standard textbooks or online references, for example, mech4study [50].

Lastly, for friction stir welding, a small rotating tool is plunged into the joint which heats and stirs the materials [51, 52]. Relatively, low heat input is required in order to produce minimal distortion. This process produces no

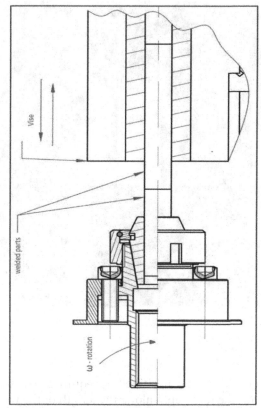

FIGURE 6.48

Rotary friction welding machine (left); Sketch of RFW rotary friction welding handle (right) [49].

FIGURE 6.49
Linear friction welding diagram [50].

fumes or spatter. However, heavy machinery is required to drive the tool into the joint and maintain the tool depth along the length of the weld. There are many research papers available in this area. Figure 6.50 shows the general principle of friction stir welding.

Figure 6.51 shows the different types of joints welded by FSW process:

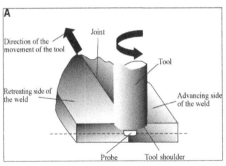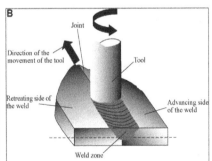

FIGURE 6.50
Schematic of the principle of friction stir welding (FSW) [51].

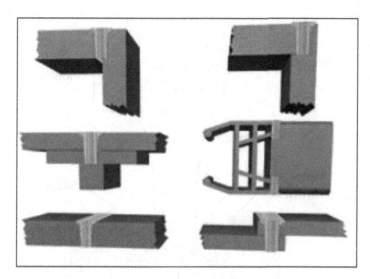

FIGURE 6.51
Examples of different types of joints welded by FSW process [52].

6.6.2.3 Ultrasonic Welding

In ultrasonic welding, frictional heat is produced by the ultrasonic waves and force is used for the joining process [52, 53]. Ultrasonic waves (15 to 60 kHz) are transferred to the material under pressure with a sonometer. Welding times are lower than 3 seconds and the welding can proceed with or without the application of external heat. The principle of the process limits the allowable mass of material on the sonotrode side to a maximum of 10 grams. The maximum thickness that can be welded depends on the self-damping characteristics of the workpiece. A main advantage while welding aluminum is the fact that the vibrations break the oxide layer and transport it to the boundary regions. As a result, mechanical and chemical surface cleaning is not necessary. Surface coatings and impurities behave in a similar manner. Consequently, the main application area for ultrasonic welding is the contact joining of wires. Schematic of ultrasonic welding principle is shown in Figure 6.52 [52]:

Ultrasonic welding is particularly suitable for joining aluminum and its alloys with each other as well as producing composite joints with other materials. In electrical and electronic applications, the frictional energy creates clean welding zones with low contact resistances. Hardness influences the weldability of composite joints with steel. Hard alloys are less suitable since the plastic formability required for joining is not sufficient. Very often an intermediate layer is used to overcome this deficiency.

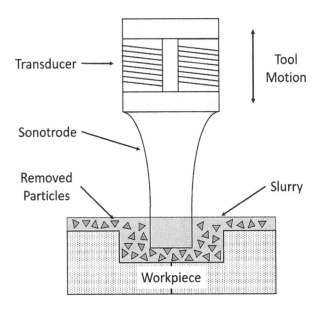

FIGURE 6.52
Principle of ultrasonic welding.

6.6.2.4 Explosion Welding

Explosion welding topic was also covered earlier (Figures 6.9 and 6.10). It is a process where two metals are stacked on top of each other followed by a layer of explosive powder being placed on top of the plates. Upon detonating the explosives, a huge amount of downward force is applied to the materials, thus welding them together. The process creates a wavy interface between the metals, increasing the shear strength of the bond. The process is typically used to weld together traditionally incompatible materials. The shock waves can produce pressures of up to 60 GPa.

6.6.2.5 Mechanical Joining Processes

This topic has also been discussed before from a different perspective. Until recently, mechanical joining was the main technology used to join aluminum and steel components. In general, all the different mechanical joining methods used within the automotive industry are also suitable to join dissimilar metals [54]. However, depending on the material combination (strength and ductility of both partners), some limitations may exist in particular for mechanical joining techniques which are based on forming and cutting processes (i.e., clinching, self-piercing riveting, flow-drilling screws, etc.). In most cases, mechanical joining techniques are combined with adhesive bonding to increase the static and fatigue strength of joints and prevent deterioration of corrosion resistance of joints caused by contact between dissimilar metals.

Hem flange bonding is the standard solution when aluminum and steel closure panels are joined. In order to prevent from problems related to galvanic corrosion and thermal deformation between an aluminum outer panel and a steel inner panel, an adhesive is used. Figure 6.53 shows the principle of this process.

6.6.3 Joining Aluminum to Plastics

Joining aluminum to plastics is a challenging area. Adhesive bonding is the least expensive joining method for joining aluminum to plastics permanently but is not necessarily the best solution. For assemblies that must be taken apart a limited number of times, mechanical fasteners are the least expensive. However, if an assembly is intended to be taken apart regularly, metal inserts in the plastics are considered. Rivets also offer a simple, easily automated installation process that can be used in particular for plastic to sheet metal joints.

Laser-assisted metal and plastic joining is applicable to many combinations of metals (i.e., steels, titanium, and aluminum alloys) and plastics (i.e., PET, polyamide (PA), and polycarbonate (PC)) [55]. As shown in Figure 6.53, the laser beam heats the metal either from the plastic of the metal side of a lap joint and metals the plastic near the joint interface. The key point is the formation of small bubbles (diameter of 0.5 mm or less) which induce a high pressure in the molten plastic. This molten plastic is forced to the metal surface. Anchoring

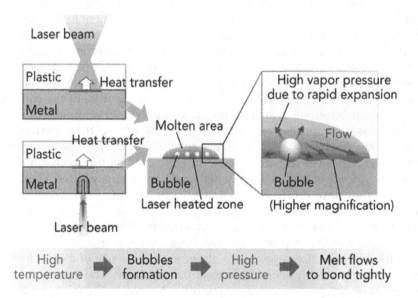

FIGURE 6.53
Mechanism of laser-assisted metal and plastic joining [55].

effects in concavities of the surface topography, physical Van der Waals forces, and chemical bonding through the oxide film produce a strong joint. Some of these topics are discussed in detail in references [55] and [56].

6.6.4 Joining Aluminum to Composites

Composite materials consist of a polymeric matrix resin which is used to bind fibrous reinforcements into the combined material. The type, volume fraction, length, and layup of the fibers determine its mechanical properties. Composites can have complex directional mechanical properties differing by inplane and outplane directions. In the fiber direction, properties, such as tensile strength, Young's modulus, and yield strength are considerably higher and depend on adhesion between the resin matrix and the fiber. Perpendicular to the fiber, the properties approach those for the matrix resin alone.

Joints between aluminum and either a thermoset or thermoplastic composite are generally achieved by adhesive bonding or a combination of adhesive bonding and mechanical fasteners. To deal with complex loading at the faying surfaces, biased fiber layers can be oriented nearest to the joint. It is also necessary to allow for a thermal mismatch. The coefficient of thermal expansion for epoxy-based parts matches that of aluminum fairly well. In comparison, the coefficient of thermal expansion for a thermoplastic composite, such as glass fiber polypropylene, is higher, and so has the potential for thermal strains at the bond line.

Transition joints are important auxiliary means to join metals and composite materials. They are normally metallic elements which are integrated into the composite during part manufacturing. Attachment to the metal part is then possible using conventional joining methods. Different concepts are evaluated to realize transition structures between aluminum and carbon fiber-reinforced composites. As an example, using the Stir-lock technique, reinforced transition joints can be produced. In another example, a stainless steel mesh, which provides a skeleton for the application of the composite material can be joined to an aluminum element by friction welding [56].

Thermoset composites are easier to adhesively bond than thermoplastic composites because they have higher surface wettability. Epoxy, urethane, or acrylic adhesives can all be used for adhesive bonding with aluminum. Epoxies are especially reliable when used with epoxy-based composites because they have similar flow characteristics. Thermoplastic composites do not wet well with adhesive and, usually, require a form of surface activation. This can be a flame, corona, or a plasma treatment that oxidizes the surface to increase wetting. A primer can also improve wetting. Another possibility to adhesively bond composites to aluminum is the use of a primered aluminum component (preferably an electrocoat primer). The bond would then be between the primer and the composite. Normally, the joint can then be designed with the assumption that the weak link is the primer/aluminum interface.

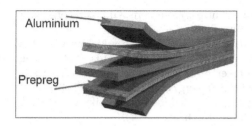
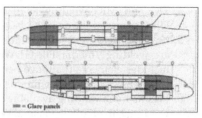

FIGURE 6.54
A component view of a GLARE3-3/2 hybrid sheet (left); Areas of the airbus 380 aircraft fuse-lage where the glass-laminated aluminum-reinforced epoxy (GLARE) structural material is applied (right) [57].

Mechanical fasteners are often used to join aluminum with plastics and composites, in many cases, combined with adhesives. Different standard mechanical joining methods (i.e., rivets, two-piece bolts, or blind fasteners) use predrilled holes in both the aluminum and composite parts. When specifying the applicable mechanical fasteners, several factors must be considered. Thermal expansion of the fastener and the joined materials, the effect of drilling on the structural integrity of the component as well as the possible fiber delamination caused by a loaded fastener, the possibility of water (humidity) intrusion between the fastener and the aluminum/composite material, and possible galvanic corrosion effects are all important factors to consider.

Fasteners for composites should have large heads to distribute the load over a larger surface area in order to reduce crushing of the composite material. Interference fits may cause delamination of the composite and should be avoided. If an interference fit is necessary, special sleeved fasteners can limit the chances of damage in the clearance hole. Fasteners can also be bonded in place with adhesives to reduce fretting. Drilling and machining damage composite materials. The number and size of defects (delamination, resin erosion, or fiber breakout) allowed in a structure depend on the application.

As an application in aerospace industry, a component view of a GLARE3-3/2 hybrid sheet can be seen in Figure 6.54. There are three layers of aluminum alternating with two layers of glass fiber. In a GLARE3 grade, each glass fiber layer has two plies: one oriented at zero degrees, and the other oriented at ninety degrees [57].

References

1. European Aluminum Association and MATTER (C) 2001-2010 MATTER.
2. Ostermann, Friedrich TALAT – A Training Program for Aluminum Application Technologies in Europe, Materials and Engineering A199.1 1995: 73–77.
3. Hajli, Samir. "Advanced Manufacturing Welding Processes for Joining Dissimilar Metals and Plastics," Technical Report, Politecnico Milano.

4. TIG and MIG welding available at: https://commons.wikimedia.org/wiki/File:Gas_arc_welding_(TIG_%26_MIG).PNG
5. Gas arc welding available at: https://en.wikipedia.org/wiki/Gas_tungsten_arc_welding#/media/File:GTAW.svg Duk – Own work This vector image includes elements that have been taken or adapted from this file: GTAW.png., CC BY-SA 3.0, https://commons.wikimedia.org/w/index.php?curid=455575
6. Plasma torch welding available at: https://en.wikipedia.org/wiki/Plasma_arc_welding#/media/File:Plasma_Welding_Torch.svg
7. https://commons.wikimedia.org/wiki/File:Spot_welding_process.png
8. T.S. Hong, Morteza Ghobakhloo and Weria Khaksar. (2014). "Robotic Welding Technology." Comprehensive Materials Processing. Elsevier.
9. Resistance seam welding process available at: https://techminy.com/resistance-seam-welding/
10. Encyclopedia or welding terminology available at: https://www.weldcor.ca/index.php/encyclopedia.html?alpha==B&per_page=5
11. Friction welding process available at: https://techminy.com/friction-welding/
12. Friction welding process available at: https://www.twi-global.com/technical-knowledge/job-knowledge/linear-friction-welding-146
13. Explosive welding process available at: https://en.wikipedia.org/wiki/Explosion_welding#/media/File:Explosion_welding.png
14. Explosive welding process available at: https://mechanical-engg.com/gallery/image/790-explosive-welding/
15. Electromagnetic forming available at: https://commons.wikimedia.org/wiki/File:Electromagnetic_Forming_01.png
16. Principle of ultrasonic-welding available at: https://commons.wikimedia.org/wiki/File:Ultrasonic_Welding.JPG
17. U. Kruger. "TALAT-Friction, Explosive and Ultrasonic Welding Processes of Aluminum," chap. 4400. EAA: European Aluminum Association, 1994. TALAT: Training in Aluminum Application Technologies.
18. Mechanical fastening available at: https://www.sciencedirect.com/topics/materials-science/mechanical-fastening
19. Mechanical joining available at:https://www.swantec.com/technology/mechanical-joining/
20. E. Patrick and M.L. Sharp Joining Methods for Aluminum Car Body Structures, Automotive Technology International (1993), 61–70.
21. Simulation of hemming process available at: https://www.autoform.com/en/glossary/hemming/
22. Xiaocong He, Clinching for Sheet Materials, Science and Technology of Advanced Materials 18(1) 2017: 381–405, Taylor & Francis (Open Access)
23. Clinching Process by Nestor available at https://commons.wikimedia.org/wiki/File:Clinchar.png
24. Different phases of Clinching Process by Francesco Lambiase, available at: https://commons.wikimedia.org/wiki/File:G5151Clinching_Phases_from_Lambiase_%22Clinch_joining_of_heat-treatable_aluminum_AA6082-T6_alloy_under_warm_conditions%22.png
25. Different Clinching Dies by Francesco Lambiase, available at: https://upload.wikimedia.org/wikipedia/commons/4/4d/Clinching_Dies_from_Lambiase_-_%22Mechanical_behaviour_of_polymer-metal_hybrid_joints_produced_by_clinching_with_different_types_of_dies%22.png

26. Bolted Joint by Yuri Raysper available at: https://commons.wikimedia.org/wiki/File:Bolted_joint.svg
27. http://alcoainnovation.com/fr/pdf/Donald_Spinella-Joining_Methods_nov21.pdf
28. A self-tapping threaded insert by Sjgroovpin, available at: https://commons.wikimedia.org/wiki/File:Taplok_Insert.jpg
29. Self-tapping screws by Jonathan Gibert, available at: https://commons.wikimedia.org/wiki/File:Cataphoresis.JPG
30. Simple concept of thread forming by "wizard191" available at: https://commons.wikimedia.org/wiki/File:Thread_forming_and_rolling_concept.svg
31. Solid rivets by Miaow and Miaow, available at: https://commons.wikimedia.org/wiki/File:Rivet01.jpg
32. Example of a 4-step orbital rivet by Orbitform, and by Dgshirkey, available at: https://en.wikipedia.org/wiki/Riveting_machine, and https://commons.wikimedia.org/wiki/File:Orbital_4-step_best_copy.jpg
33. Blind rivet notches by Sarang, available at: https://commons.wikimedia.org/wiki/File:Blind_rivet_notches.jpg
34. Example of lockbolts, available at: https://www.afshuck.net/es/sobre_nosotros/.html
35. Self-piercing rived (SPR) joint by R. Haque et al., "SPR Characteristics Curve and Distribution of Residual Stress in Self-Piercing Riveted Joints of Steel Sheets", Advances in Materials Science and Engineering, vol. 2017, Article ID 5824171, 14 pages, 2017. https://doi.org/10.1155/2017/5824171
36. Clinch riveting available at: https://us.tox-pressotechnik.com/terms-of-use/
37. Example of Tack high-speed joining available at:
    a) https://www.boellhoff.com/de-en/products-and-services/assembly-technology/high-speed-joining-rivtac.php
    b) RIVTAC_Automation_P_Titel_DSCF7810.zip   https://bollhoffautomation.egnyte.com/dl/y1uNPlRFoh
    c) Rivtac Setzfolge.tiff https://bollhoffautomation.egnyte.com/dl/gzd0egWc15
    d) Rivtac Detailansicht.tiff https://bollhoffautomation.egnyte.com/dl/RdzesIVPea
38. Typical front wheel drive system, available at: https://commons.wikimedia.org/wiki/File:1965_Austin_Mini,_Sectioned_Heritage_Motor_Centre,_Gaydon.jpg) - Mark Brown from Hampton, New Brunswick, Canada
39. Layers of Aluminum, available at: https://www.european-aluminium.eu/
40. Failure of adhesive joints by Fraunhofer ENAS, D. Neumann, Wiemar, et al, available at: https://commons.wikimedia.org/wiki/File:B-ad-schematicsu8bondingprocess.png
41. Failure of adhesive joints by Rswarbrick, available at: https://commons.wikimedia.org/wiki/File:AdhesiveFractures.svg
42. https://www.european-aluminum.eu/media/1524/9-adhesive-bonding_2015.pdf https://www.antala.uk/surface-preparation-for-adhesive-bonding/
43. Surface treatment available at: https://www.adhesives.org/adhesives-sealants/adhesives-sealants-overview/use-of-adhesives/surface-treatment/surface-pretreatment
44. Rehua, A common example of a galvanic cell, available at: https://commons.wikimedia.org/wiki/File:Galvanic_cell_labeled.svg
45. (a) StructX.com. "Galvanic Series (Scale of Nobility)." *Galvanic Series (Electrochemical Series)*, structx.com/Material_Properties_001.html

(b) Tugsataydin, Galvanic series, 22 November 2017, available at: https://commons.wikimedia.org/wiki/File:Galvanic_series_noble_metals.jpg

46. Peter Lewes, Galvanic corrosion in the Statue of Liberty, available at: https://commons.wikimedia.org/wiki/File:Statue_lib_corr1.png

47. Torsten Henning, "Electroplating." *Wikipedia*, Wikimedia Foundation, 4 March 2019, en.wikipedia.org/wiki/Electroplating b) Chetan, 18 October 2005, https://en.wikipedia.org/wiki/Galvanic_corrosion

48. Cafe Nervosa, "Simple impressed current cathode protection system, 7 June 2012, available at: https://en.wikipedia.org/wiki/Cathodic_protection. U.S. Government, "The 'light patches' are zinc blocks used as sacrificial anodes," 23 May 2007, available at: https://commons.wikimedia.org/wiki/File:Ship-propeller.jpg

49. (a) Rotary Friction Welding Machine by TFW, available at: https://en.wikipedia.org/wiki/File:Rotary_friction_weld.jpg

    (b) Sketch of RFW Rotary friction welding handle by Xyz00030280, available at: https://commons.wikimedia.org/wiki/File:RFW_Rotary_friction_welding_handle.jpg

50. "Friction Welding: Principle, Working, Types, Application, Advantages and Disadvantages, available at: https://www.mech4study.com/2017/04/friction-welding-principle-working-types-application-advantages-and-disadvantages.html

51. Examples of different types of joints welded by FSW process, Nova-Tech Engineering, available at: https://commons.wikimedia.org/wiki/File:Shapes_and_joints.png

52. Principle of ultrasonic welding by Four30, available at: https://commons.wikimedia.org/wiki/File:Ultrasonic_Machine_Process.jpg

53. Ultrasonic welding: Sonic welding, available at: http://www.craftechcorp.com/ultrasonic-welding

54. M.R. Stoudt, et al. "Characterizing the Hemming Performance of Automotive Aluminum Alloys With High-Resolution Topographic Imaging", Figure 1, Journal of Engineering Materials and Technology JULY 2014, Vol. 136/031001-1.

55. Seiji Katayama, et al: "Laser-Assisted Metal and Plastic Joining, Proc. Of LANE 2007, pp 41-51. Article available at: https://www.industrial-lasers.com/articles/print/volume-250/issue-6/features/laser-joining-of-metal-and-plastic.html

56. W.M. Thomas, D.J. Staines, I.M. Norris, S.A. Westgate, and C.S. Wiesner. (2006). Transition joints between dissimilar materials.

57. Yang Jenn-Ming, et al. An exploded view of a GLARE3-3/2 hybrid sheet, available at: https://commons.wikimedia.org/wiki/File:An_exploded_view_of_a_GLARE_hybrid_sheet.jpg

Index

Printed in the United States
by Baker & Taylor Publisher Services